Impact Studies

Within the past two decades, research on impact cratering mainly gained attention throughout the world after the theory that a large impact event caused the extinction of ca. 50% of all living species, including the dinosaurs, approximately 65 million years ago was proposed. Impact craters are formed when a large meteorid crashes into a larger planetary body with a solid surface.

The interdisciplinary series, **Impact Studies**, aims to include all aspects of research related to impact cratering—geology, geophysics, paleontology, geochemistry, mineralogy, petrology, planetology, etc. Experimental studies on shock metamorphism, investigations of individual impact craters, comparative planetology, and paleontological research on the effects of impact events are just a few topics that are treated in this series. All contributions are peer-reviewed to ensure high scientific quality.

More information about this series at https://link.springer.com/bookseries/4698

Josep M. Trigo-Rodríguez

Asteroid Impact Risk

Impact Hazard from Asteroids and Comets

Springer

Josep M. Trigo-Rodríguez
Research Scientist
Institute of Space Sciences (CSIC-IEEC), Campus UAB
Cerdanyola del Vallés, Barcelona, Catalonia, Spain

ISSN 1612-8338
Impact Studies
ISBN 978-3-030-95123-8 ISBN 978-3-030-95124-5 (eBook)
https://doi.org/10.1007/978-3-030-95124-5

This Springer imprint is published by the registered company Springer Nature Switzerland AG
The registered company address is: Gewerbestrasse 11, 6330 Cham, Switzerland

Foreword

Meteorites represent the only material with a non-terrestrial origin that can be studied in a laboratory, with the only exception of sample-returned materials. Their study can thus give very strong clues as to the origin of the Solar system and of life itself. Knowing both the origin and indeed the life history of the meteorite is thus of great importance. The important question is where did they originate from, on comets or asteroids? Also the level of changes that have occurred from their release on the parent body until their arrival on the surface of the Earth. Some of these changes take place during the passage of the meteorite through Earth's atmosphere, while other modifications can take place while on the Earth surface, some due to interactions with humanity. Knowing the trajectory through the atmosphere and computing the orbit in the solar system are thus of vital importance to the understanding of the history of the meteorite. Trigo-Rodriguez has spent a considerable fraction of his work life on these problems and is a renowned expert in this field. In this book, he gives us the benefit of his considerable knowledge at a level that is understandable to the non-expert.

Besides understanding the history of the life of the meteorite, it is important to know what evolution takes place to the composition of the meteorite while it is still part of the parent body. This requires an understanding of the structure and evolution of the parent: For example, what changes take place following a collision involving the parent, which might result in considerable heating of both the parent and the meteorite. These aspects are also fully dealt with in the chapters of the book, also from the unique author perspective of studying these processes in meteorites and cometary materials in CSIC laboratories.

Besides being important to the advancement of science, knowing the history of a meteorite and inferring from it the physico-chemical properties of its parent body are relevant issues to the evaluation of the threat that they pose to life on Earth and on how to reduce the level of hazard. In conclusion, in this book Trigo-Rodriguez makes a considerable effort to summarize our current knowledge in these fields that will be crucial to deal with other key topics like, e.g. to be able to deflect asteroids, or to identify the most relevant targets for the exploration of our solar system in the near future.

Prof. Iwan Williams
Emeritus Professor
School of Physics and Astronomy, Queen Mary
University of London
London, UK

Introduction

Our lives are so intimately linked to planet Earth that it becomes difficult to think that the existence of humanity could suddenly be put into check. We certainly don't tend to think that a celestial phenomenon could affect us in such a way. Needless to say, this attitude denotes how a lack of wider thinking still dominates our species, even as we head towards the middle of the twenty-first century. It is not surprising in a society that still denies, at least partially, the anthropogenic influence on climate change, despite the growing cluster of scientific evidence gained in recent decades. In a similar way, science has provided us with clues about the relevance of the impact hazard over time—which is the focus of this book.

Human beings rely largely on accumulated experience and historical legacy, subject to our own interpretation of celestial phenomena that has clearly changed over time. However, our species has a relatively short written record of what has taken place, so that legacy is enormously biased. Science helps us to use rocks to decipher evidence of certain phenomena that once occurred on our planet, as inexorable as they were fearsome, and that knowledge may help us to change the way we think. In this book, I intend to disseminate the teachings of various disciplines, some sadly relegated, regarding impacts with asteroids and comets. Astrophysics, geology, geophysics and palaeontology come together in a research field quite unknown to the general public. This book in your hands is intended to circumvent this lack of rigorous information, by spreading the knowledge of the role that hecatombs have played in the history of life on Earth.

Despite the relative rarity of impacts in our collective imagination, asteroids and comets are common objects that many scientists passionately study, and they are often the focus of scientific news stories. On occasions, some of these objects have been the protagonists of close encounters with our planet that incite us to reflection. However, the written press is very often more concerned with sensationalism than rigorous disclosure. All this trivialises a very serious problem and, as a direct consequence, this danger coming from the heavens does not receive the attention it deserves. In a way it rests in our consciousness, and seems to only awaken and activate our interest if we hear news about the dangerous approach of an asteroid to Earth. Should we think these are just something to behold in the context of science fiction? What if, on a given day, an unexpected encounter changes our perception of the problem (as in fact the Chelyabinsk event did)? Why do we not get down to business if this is probably the greatest avoidable danger we face?

The geological record indicates that impacts of celestial bodies have been extremely common in Earth's history, although they have been declined exponentially over time. Fortunately, today an impact happens on average on a timescale of centuries for bodies larger than 10 m in diameter. The *Center for Near-Earth Object Studies (CNEOS)* maintains a database of the largest fireball explosions detected on the top of the atmosphere using satellite sensors and other techniques, which also helps to make visible this continuous and unexpected threat.

The geological record provides indisputable evidence about the role of catastrophic impacts that have marked the evolution of life on Earth. Until a few decades ago, we were unaware of the protagonists of that danger, but in recent times we have completed the cataloguing of the largest potentially hazardous asteroids. The great powers have become aware of this latent danger and have launched monitoring programmes that, in the coming decades, will possibly give way to palliative space projects. This is not the time to take direct action against any specific object, because we have no evidence of a short-term impact, but it is time to stand guard, to complete cataloguing and tracking of potentially dangerous objects, to implement exploration missions and return samples so that we better understand the danger we face and, above all, start being prepared.

Society has slowly become aware of asteroids and comets as sources of hazard. To popularise this problem, many scientists have taken a key role in spreading their concerns and coordinating their efforts as the *Asteroid Day* initiative. The local destruction caused by the shock wave associated with the Chelyabinsk superbolide was a staging of the danger associated with impacts that served as the perfect anteroom for such an initiative. There is not only danger when a direct impact occurs but also when digging a crater or generating a tsunami of gigantic dimensions. Large fireballs, exemplified by the 30 June 1908 event on Tunguska, can become efficiently fragmented and, even if they do not generate a crater, can crush an entire region with their shock wave just a few minutes after incineration upon arrival of the hot gas that results from the ablation of the rock that produces it. A remote event over the remote Tunguska area, also far away in time, is no longer the only contemporary record. The magnitude and danger of the huge superbolide that appeared on Chelyabinsk was a reminder that impacts of asteroids of a few dozen meters, such as the one on Tunguska in 1908, may be more frequent than we thought and could happen once a century.

Thanks to multiple filming and the subsequent scientific studies on the Chelyabinsk bolide, the meteoroid origin of the earlier Tunguska explosion has been recognised, leaving behind other baseless theories. We must face up to this real and common danger to all humanity that should make us think about what we do to defend ourselves, not from our fellow men, but from that abstract danger that should unite us for a common goal of self-defence. I do not intend to fall into utopias, but necessity can lead us to an end of multinational collaboration in which we are not too accustomed. Tackling this danger will mean not only earning points towards the survival of our species, but will also bring progress towards the development of new scientific-technological disciplines, such as manned exploration of other worlds and space mining.

Discovery, monitoring and cataloguing programmes of so-called smaller bodies have been collecting information for several decades. We know quite well the danger associated with the largest potentially hazardous asteroids, those whose diameter ranges from a hundred metres to a few kilometres, but we still do not know a good deal about those of a few tens of metres, such as the one that starred in the meteoric fall of Chelyabinsk. Our shortcomings continue to be a concern that motivates the emergence of new telescopic tracking programmes and we even hope that a project for the use of space telescopes dedicated to completing the cataloguing and tracking of these objects will soon emerge.

But our concern is not only asteroids, comets should also be considered a source of hazard, although, as we will see in this book, their frequency of impact is an order of magnitude lower than that associated with asteroids. Although the danger may be statistically less, it is not at all negligible with the arrival of unexpected long-term comets in mind. As if that evidence were not enough, on 10 October 2015, with the discovery of an object more than 600 m in diameter called 2015 TB145, we became collectively aware of the existence of extinct comet nuclei with troubling peculiarities: dark, fast and unexpected. That object was discovered with only 3 weeks of margin and fortunately passed at a distance somewhat greater than that of the Moon. All of this makes us wonder whether the random nature of these encounters capable of ravaging our planet should not provide us with added motivation to develop palliative systems to combat these threats. In addition, it also makes us wonder whether, to ensure the persistence of our species, we should colonise the Moon and Mars.

This book also aims to describe the nature of the bodies that endanger us, and exemplify some of the experiments we are performing to understand their challenging physical properties—from the rocky matter, with varying degrees of hardness and density that makes up asteroids, to the mixed and porous materials we attribute to comets. The latter are formed by a mixture of rock particles, ices and organic matter, extremely rich in volatile elements whose burst into the atmosphere can become a harmful chemical bomb, as demonstrated by the studies made by *Stardust* (NASA) and *Rosetta* (ESA) spacecrafts. Several generations of astronomers have been following and determining the orbits of asteroids and comets during the last centuries, and telescopic surveys have just started to discover interstellar objects. Now our modern computers integrate the orbits of these elusive bodies to study how they evolve over time, trying to match asteroid complexes as fragments of cometary disruptions. Comet 2P/Encke and the Taurid complex is an example of the real nature of that hazard: a hydra with many heads. Such studies allow us to understand the dynamic mechanisms that bring these bodies closer to Earth. Based on these approaches, and reinterpreting the geological record, some essential data on the frequency of impacts against our planet can be obtained.

The relevance of impact hazard is clearly demonstrated as one of the biggest challenges for the survival of human beings at mid-term. Geological studies demonstrate that the stratigraphic legacy harbours clear evidence of the transcendence of these rare encounters in the history of life on our planet. The study and quantification of previous catastrophes give us clues to be able to face future survival challenges in the face of possible impacts.

It would be illogical to think that Earth's interaction with its environment is reduced to large impacts. Every night, Earth intercepts millions of particles, and tens of meteor falls occur around the globe every year. Most meteoroids reaching our planet are tiny particles, from tens of microns to a few centimetres, which evaporate to a large extent but provide materials to the high atmosphere and condense producing microspherules that gravitationally sediment until they reach the ground. The study of lake and ocean sediments reveals the magnitude of the continuous contribution of interplanetary matter reaching Earth at the rate of about 200,000 tonnes per year. In addition to this continuous and intense flow of tiny particles, dozens of meteorite drops are produced each year announced by spectacular fireballs. Such falls allow us to study the formative materials of asteroids in our laboratories, being the best proxies to the substance of these asteroids and comets.

Therefore, with this book I intend to emphasise the need to become aware of the danger associated with asteroid and comet impacts. I provide an update of our current knowledge on these challenging objects, addressing the main sources of hazard to life on Earth based on recent scientific evidence, with no intention of creating alarmism. I will emphasise that, if we really want to quantify and explain such hazards properly, we must encourage space missions because the in situ study of these minor bodies is the only way to learn about the challenging physico-chemical properties of these objects. We can identify a dense metallic or rocky asteroid as a dangerous projectile, but a more fragile hydrated asteroid or comet with enough mass to produce a Tunguska event could transform into a chemical bomb with ugly consequences for the biosphere. We will see that, although they are fortunately rare events, our written history contains evidence of fire arrived from the sky.

The exploration of undifferentiated and water-rich transitional asteroids, well exemplified by the new missions *OSIRIS-REx* (NASA) and *Hayabusa 2* (JAXA) addressed to 101955 Bennu and 162173 Ryugu, respectively, have the added value of being able to answer key questions about their nature. The challenging physical properties of these survivors of colossal impacts require all our efforts, and will perhaps provide great clues to tackle space mining as the next step.

After all, we know we can identify ourselves as the fruit of the opportunity that an asteroid provided to mammals about 65 million years ago to evolve after an impact-induced hecatomb that marked the decline of dinosaurs. If science has allowed us to learn from the geological book of the Earth's past, we should consider ourselves a fortunate species and use these unique teachings to stay alert and be ready to deal with this real problem.

Summary

1. Fireballs announcing meteorite falls.
2. Parent bodies of meteorites and their dynamic pathways to Earth.
3. The origin and nature of comets.
4. The impact associated with short-period comets and their origin.
5. Quantifying the hazard associated with asteroid impacts.
6. Impact craters and extinctions in the geologic record.

Barcelona, Catalonia, Spain
November 2021

Josep M. Trigo-Rodríguez, Ph.D.

Contents

1 Fireballs Announcing Meteorite Falls ... 1
1.1 How Many Tonnes Reach the Earth? ... 5
1.2 How is a Meteor Created? Ablation in The Atmosphere ... 5
1.3 The Search for Meteorites ... 9
1.4 The Value of Recovering Meteorites ... 14
References ... 18

2 Meteorite Parent Bodies and Their Routes to Earth ... 21
2.1 The Origin of Asteroids and Meteorites ... 22
2.2 The Main Asteroid Belt ... 23
2.3 Resonances as Sources of Meteorites ... 32
2.4 Transitional Objects and Comets ... 34
2.5 Comets: The Trans-Neptunian Belt and the Oort Cloud ... 35
References ... 39

3 The Origin and Nature of Comets ... 43
3.1 Impact with Comets ... 48
3.2 The Stardust Mission and the 81p/Wild 2 Materials ... 50
3.3 The Rosetta Mission and Comet 67P/Churyumov–Gerasimenko ... 54
3.4 The Trans-Neptunian Belt ... 57
References ... 59

4 The Risk Associated with Short-Period Comets and Its Origin ... 61
4.1 The Fragile, Dark and Elusive Nature of Comets ... 62
4.2 Impacts, Atmospheric Bursts and Their Markers ... 63
4.3 Are These Atmospheric Airbursts Recorded in History? ... 66
4.4 Asteroids and Comets in the Near-Earth Region ... 67
4.5 Comet Encke and the Taurid Complex ... 68
4.6 Was the Tunguska Event a Warning from Comet Encke? ... 70
4.7 A Great Pumpkin for Halloween: 2015 TB145 ... 72
References ... 76

5 Quantifying the Nature and the Magnitude of the Hazard from Asteroid Impacts ... 79
5.1 Shaping the Sources of Hazard ... 80
5.2 Monitoring Programmes ... 84
5.3 Asteroids Subject to Capricious Forces ... 88
5.4 Spinning-Top Asteroids ... 90
5.5 The Tidal Effect in Near-Earth Passages ... 92
5.6 Dating the Stay of Meteorites in Space ... 93
5.7 The Spaceguard Asteroid Discovery Milestone ... 96
5.8 The Current Danger of Impact by Small Asteroids ... 98
5.9 What Lessons Should We Learn? ... 100
References ... 104

6 Craters and Extinctions in the Geological Record ... 107
6.1 The Geological Scale: Looking for Traces of Catastrophes ... 111
6.2 Phanerozoic Extinctions ... 113
6.3 The Forerunners of Impact Theory ... 114
6.4 The Recognition of the Chicxulub Catastrophe and the k/t Boundary ... 116
6.5 The Impact of Asteroid Eltanin ... 119
References ... 125

Chapter 1
Fireballs Announcing Meteorite Falls

The night sky seems immutable and in the short space of a human lifetime the stars appear eternally static in their positions relative to each other. However, our static vision of the universe is just a mirage, a mere appearance that, under a starry night, the meteoric phenomenon is capable of dismantling in a few minutes. Every night, dozens of shooting stars streak across our firmament at a distance that, although it seemed stellar to our ancestors, takes place in the Earth's atmosphere, generally less than a hundred kilometres above our heads.

These rapid streaks of light exemplify that interplanetary space is far from empty, but is littered with dust and rocks that are the direct product of the erosion of asteroids, comets and the planets themselves. Although we may not be aware of it, through collisions between asteroids or the sublimation of the ices that form comets, small particles or even large rocks are propelled into the interplanetary medium at supersonic speeds. Such fragments travel at relative speeds of hundreds of thousands of kilometres per hour. When intercepted by the Earth, the particles hit the gases in the atmosphere, creating the light that produces the shooting star.

When the Earth intercepts the swarms of particles left behind by asteroids and comets, they produce what are known as shooting star showers. Each meteor follows a different trajectory among the stars and, if we were to trace them, we would see that every night there are regions of the sky that are sources of shooting stars. Such areas of sky are called radiants (Fig. 1.1).

Returning to the luminosity of stars and meteors, astronomers have established a logarithmic system to define the magnitude of brightness of the stars we look at. Thus, in astronomical convention, the most luminous stars have negative magnitudes: the most luminous star is Sirius which has magnitude -1.5, while Venus usually has magnitude -4. There are about 6,000 stars visible to the naked eye that have magnitude $+6$ or lower. The concept of magnitude stems from the pioneering work of Hipparchus of Alexandria (?–127 BC), who divided the visible stars into six classes with the brightest having magnitude 0 and the faintest visible to the naked eye having magnitude $+6$. Based on this classification, it was Norman R. Pogson (1829–1891) who, using this division into magnitudes, redefined them in an attempt

J. M. Trigo-Rodríguez, *Asteroid Impact Risk*, Impact Studies,
https://doi.org/10.1007/978-3-030-95124-5_1

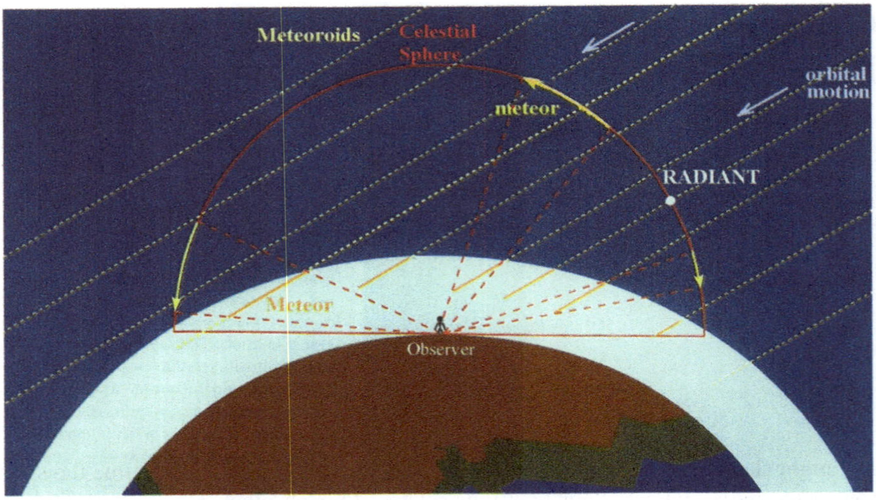

Fig. 1.1 The radiant of a meteor shower is produced by the perspective in which the particles strike the atmosphere from a meteor swarm where the particles follow similar orbits (J.M. Trigo/CSIC-IEEC)

to adapt Hipparchus' work. Since Hipparchus of Alexandria had underestimated the magnitudes of the brightest stars, this led to some stars later having negative magnitudes.

We will use the stellar magnitude system to estimate the magnitude of meteors. On a starry night we can see dozens of meteors, the most frequent being those of similar magnitude to the faintest stars (Jenniskens, 2006; Trigo-Rodriguez, 2019). On such nights we can also see quite a few meteors as bright as the bright stars, and even a few whose luminosity is equal to or greater than that of the planet Venus, which are known as fireballs. The meteors become more infrequent as their luminosity increases. However, on certain nights when the Earth passes through the solid remains of a comet, we may be surprised by the appearance of bright fireballs in uncommon profusion (Fig. 1.2). Because of the spectacular and unpredictable nature of these celestial phenomena, there are many amateurs who follow and study them in a similar way to how astronomers and popularisers of the stature of Camille Flammarion (1842–1925) and Josep Comas i Solà (1868–1937) did in their day.

As we can imagine, meteoroids of tens of micrometres or a few millimetres will generate meteors in the visible range, while those of centimetre scale will originate fireballs. The resulting magnitude of meteor brightness is essentially proportional to the kinetic energy of entry into the atmosphere as the ionised plume. At a certain depth of penetration, the collision with atoms in the atmosphere superficially heats the meteoroids until they melt, fragment and evaporate, giving rise to glowing fireballs. Sometimes the fireballs can rival the luminosity of the Moon itself, and the luminous phase of the ionised plume propagates into the lower parts of the Earth's atmosphere. In fact, the reconstruction of the luminous trajectories of the most luminous fireballs

Fig. 1.2 The Earth crosses cometary dust trails, and the nights are plagued by meteors and fireballs. Circumpolar photograph showing a fireball from the Perseids, 1993 (J.M. Trigo/CSIC-IEEC)

allows us to recover *meteorites*, surviving rocks from the farthest reaches of our planetary system (Trigo-Rodriguez et al., 2006). Table 1.1 shows the main cases of meteoric phenomena.

Terminology : Asteroid, Comet and Meteoroid

As we have seen, the constant appearance of shooting stars exemplifies that interplanetary space is far from empty. These particles streaking through interplanetary space are scientifically known as *meteoroids*. The International Astronomical Union (IAU) defined a *meteoroid* as a solid object larger than an atom or molecule and with a maximum diameter estimated to be about 1 *m*. Above that diameter we speak of *asteroids* if they are rocky, rocky-metal or metallic objects, while we speak about *comets* when they are pristine objects consisting of minerals, ices and organic matter. The latter usually develop gaseous envelopes and tails as a consequence of the sublimation of ices and volatile organic compounds.

Often the terminology is confused in newspaper reports, for example when discussing the coming of the next meteorite when they should be talking about an asteroid. Meteorites are the rocks that survive atmospheric ablation to arrive on the terrestrial surface. They come from much larger rocks, given that they lose more than 95% of their mass. Those rocks no larger than one metre are called *meteoroids* while in interplanetary space. They have been dislodged

Table 1.1 Classification of meteoric phenomena according to their magnitude

Title	Magnitude (m)	Examples and main features
Micrometeor	$m < +6$	Visible through binoculars or telescopes. Evidence of the continuous inflow of interplanetary dust particles into the Earth's atmosphere
Meteor (Popular synonym: *shooting star*)	$+6 < m < -4$	Dozens are observed on a starry night. They are brief, typically lasting between two and five seconds
Bolide (Popular synonym: *fireball*)	$-4 < m < -17$	Spectacular but brief phenomena. Sometimes accompanied by persistent trails for a few minutes and even by sound. They are produced by particles from a few centimetres to several metres in diameter. The most luminous ones may survive their atmospheric transit and reach the ground as *meteorites*
Superbolide	$m \geq -17$	Very rare phenomena that often produce large meteorite falls, such as the Chelyabinsk superbolide

from asteroids or planetary bodies, such as the Moon, Vesta or Mars, by large collisions. Today we know that rocks of only a few metres in size pose little danger as they usually fragment in the atmosphere. In fact, in the entire written history of mankind, there is no known death caused by a direct meteorite impact (Trigo-Rodriguez, 2012a, b). However, these phenomena are truly spectacular given that the rocks that streak through the solar system encounter the Earth at speeds that are enormous by our standards. Meteoroids strike the Earth's atmosphere at speeds ranging from 11 to 42 km/s. The lower limit is given by a falling meteoroid under the Earth's gravitational pull and the upper limit is defined by the escape velocity of the solar pull. Since the Earth has an orbital velocity of about 30 km/s, a variety of geometries can occur in the encounter with these particles, from a meteoroid that reaches the Earth with the minimum velocity to a meteoroid that reaches the Earth head-on (42 + 30 = 72 km/s). Therefore, meteoroids collide with geocentric velocities between 11 and 72 km/s, or between about 40,000 and 260,000 km/h (Trigo-Rodriguez, 2019). Studying meteors reveals that many of them do not travel alone. Comets whose orbit crosses the Earth's orbit are often the source of meteor swarms, true complexes of fragments of different dimensions that cross the solar system in similar orbits. When the Earth crosses these swarms, the well-known meteor showers occur.

We have seen that the velocities of meteoroid encounters with the Earth are variable, but that certain average values can be defined as the most likely. Statistically, the average velocity of impacts with fragments of Near-Earth Asteroids (NEAs) is estimated to be 16 km/s, while for members of the Trans-Neptunian belt it would be around 20 km/s and around 24 km/s for asteroids from the main belt (Trigo-Rodriguez, 2012a). Long-period comets can already have much higher encounter velocities. Therefore, the meteoroids have many different sources.

1.1 How Many Tonnes Reach the Earth?

We might think that the magnitude of this meteoric flux is small, but we would be wrong. The quantities speak for themselves: estimates by various methods confirm that between 40,000 and 80,000 tonnes of interplanetary matter reach the surface of our planet every year—a not inconsiderable amount (McSween, 1999; Brownlee, 2001). Most of this material enters the atmosphere as micrometre-sized particles that give rise to what is called the sporadic component. Our estimates show that most of the total flux of cometary materials reaching the Earth comes from the sporadic background, while the major meteoroid streams deliver few thousand tons per year (Trigo-Rodríguez and Blum, 2021). This enormous amount of particles produces the so-called Zodiacal Light by reflecting sunlight in those regions of the ecliptic plane that is defined by the Earth's orbit around the Sun. The zodiacal dust particles represent materials shed from comets or asteroids tens of thousands of years ago.

Comets disintegrate very efficiently during their perihelion passages, giving rise to swarms of particles that maintain their orbital affinity—i.e., their coherence and similarity to the parent comet—for thousands of years. To give an example, we recently calculated that during a meteor storm, such as the Draconid storm of 8 October 2011 associated with Comet 21P/Giacobini–Zinner, tens of tonnes of cometary material can be deposited in the atmosphere (Trigo-Rodríguez et al., 2013). Comets are the main objects that generate meteoric swarms, although some asteroids (Table 1.2) also seem to produce them (Jenniskens, 2006).

1.2 How is a Meteor Created? Ablation in The Atmosphere

Particles detached from other solar system bodies enter the atmosphere at enormous speeds. The meteor light arises from collisions between atoms, as many of them are excited or ionised and, as the electrons return to the ground state, they emit light that will give rise to the luminous columns that we call meteors. As meteoroids deepen

Table 1.2 Main meteoroid swarms, dates when meteors can be observed, position of the radiant (right ascension (RA) and declination (Dec) in degrees), the zenithal hourly rate (THZ) or maximum number of meteors/hour under perfect sky quality conditions and radiant at zenith. Note that an asterisk indicates that it may undergo outbursts, associated with dense curtains of material, in which the meteor activity is several times higher than indicated. The last column indicates the parent bodies of the above swarms (if known). NEO is the acronym for Near-Earth Object, while a number followed by P/ indicates the Minor Planet Center (MPC) catalogue number of a periodic comet (e.g., 26P/) and, finally, a C/ followed by the year and an alphanumeric code is the nomenclature used for a non-periodic comet (its name, if given, appears in parentheses)

Swarm	Observation period (maximum day)	Radiant RA (°) / Dec (°)		Geocentric velocity (km/s)	THZ (meteors/hour)	Parent body/comments
Quadrantids	1–5 January (3)	229	+49			NEO 2003EH1
Lyrids	16–26 April (21)	272	+33		30*	C/1861 G1 (Thatcher)
Pi Puppids	18–25 April (23)	110	−45		5*	26P/Grigg-Skjellerup
Eta Aquariids	21 April–12 May (5)	337		65	60*	1P/Halley
Tau Herculids	19 May–19 June (9)	229	+39		3*	73P/Schwassmann-Wachmann 3
June Bootids	27 June–5 July (27)	223	+48		5*	7P/Pons-Winnecke
Delta Aquariids	14 July–18 August	342	−15		25	It has two branches, North and South
Alpha Capricornids	15 July–11 September (August 1)	306	−9			169P/NEAT
Perseids	23 July–22 August (12)		+58	59	100*	109P/Swift-Tuttle
Aurigids	25 August–6 September	85	+41		10*	C/1911 N1 (Kiess)
Draconids	6–10 October (9)	264	+58		5*	21P/Giacobini-Zinner
Orionids	15–29 October (21)	95	+16		30*	1P/Halley
Northern Taurids	12 October–2 December	58	+21			2P/Encke

(continued)

Table 1.2 (continued)

Swarm	Observation period (maximum day)	Radiant RA (°) / Dec (°)		Geocentric velocity (km/s)	THZ (meteors/hour)	Parent body/comments
Quadrantids	1–5 January (3)	229	+49			NEO 2003EH1
Southern Taurids	17 September–27 November	49	+13	58	+21	NEO 2004 TG10
Leonids	14–20 November (17)		+22	71	20*	55P/Tempel-Tuttle
Alpha Monocerotids	20–23 November (21)		+1	63	10*	Unknown
Geminids	6–19 December (13–14)	113	+32	35	120*	3200 Phaeton

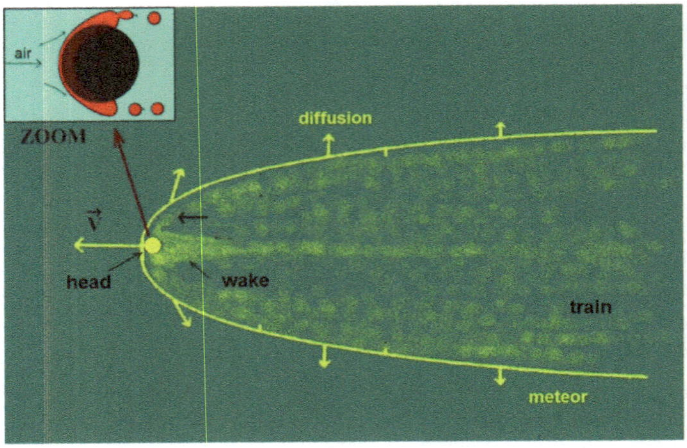

Fig. 1.3 The formation of the meteor plume by ionised material occurs during the abrupt deceleration of a meteoroid in the atmosphere (J.M. Trigo/CSIC-IEEC)

in the atmosphere, they undergo continuous collisions with molecules in the stratosphere, which gradually heat up the surface of the meteoroid until the process of mass loss known as *ablation* begins. The process begins when the surface of the meteoroids reaches temperatures of about 1500 ºC. At that moment the different constituent minerals, depending on their particular melting temperature, change phase and start to detach and evaporate to create the meteoroid (Fig. 1.3). The meteoric plumes have temperatures between 3500 and 5500 °C while the shock front can reach 10,000 °C (Borovička, 1993; Trigo-Rodríguez, 2019).

Table 1.3 shows the diameter ranges of the particles that give rise to the meteoric phenomenon. Most of the interplanetary dust that, given its low encounter velocity with the Earth and small diameter, decelerates efficiently and survives is called *interplanetary dust particles* (also known as IDPs; Fig. 1.4). Trigo-Rodriguez (2012a) defined a lower limit to distinguish between meteoroid and dust of about 100 microns (i.e., 0.1 mm). The contribution is highly variable depending on the mass range to which we refer, as shown by a graph (see Fig. 1.5) that compiles the different sources of interplanetary materials arriving annually at Earth as a function of the logarithm of their mass at atmospheric entry (Hughes, 1993).

Table 1.3 Object names, diameter ranges and associated phenomena (Trigo-Rodríguez, 2012b)

Bodies in space	Diameter range (D)	Atmospheric phenomenon	Remains
Interplanetary dust particles (IDPs)	$D < 10\ \mu m$	Progressive deposition	Micrometeorites, semi-molten spherules
Meteoroids	$0.1\ mm < D < 10\ m$	Meteors or fireballs	Meteorites (rarely craters)
Asteroids/comets	$D > 10\ m$	Superbolides	Impact craters

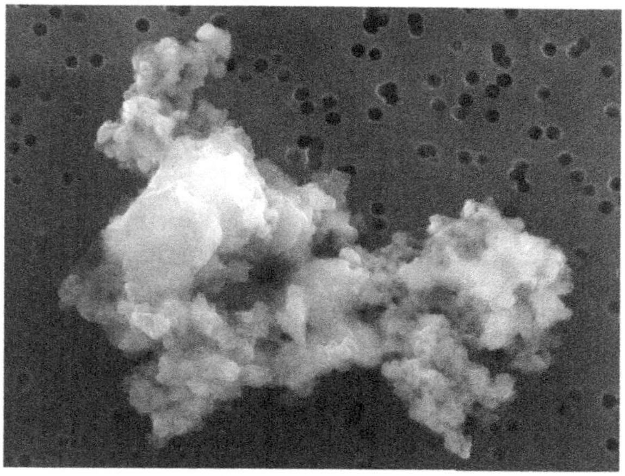

Fig. 1.4 Interplanetary dust particle or IDP. Its size is about 10 microns (courtesy of Don Brownlee/NASA)

For a better understanding, Fig. 1.6 exemplifies the nomenclature used to refer to the meteoric phenomenon in its many facets. Tiny dust particles whose dimensions usually do not exceed a few microns are efficiently decelerated and reach the Earth's surface as IDPs or micrometeorites. Micrometeorites are usually altered: partial or complete melting reduces them to glassy spherules. However, meteors are usually produced by larger particles whose masses upon entering the atmosphere range from a few milligrams to hundreds of tonnes. The smaller ones produce shooting stars or meteors, while the larger ones produce fireballs. In general, the atmosphere is an efficient protective shield, but sometimes larger rocks survive in a small percentage of their initial mass in the form of meteorites.

1.3 The Search for Meteorites

As previously explained, the *meteorites* are somehow lucky rocks. After a journey of tens of millions of years through interplanetary space, they reach the cosmic carom that will change their destiny. At one point in their orbit the Earth will get in their way and cause them to decelerate, losing most of their mass, until they reach the surface. Along this route through interplanetary space, the meteoroids will be exposed to radiation and impacts. Small rock fragments that emerge in interplanetary space as a result of the disintegration of the various bodies that populate our planetary system naturally survive as meteorites. Recovering meteorites is therefore a valuable scientific opportunity in many respects.

These extraterrestrial rocks are falling continuously across the globe. However, current searches concentrate on both hot and cold (icy) desert regions. Over thousands of years, meteorites have fallen in these regions where they accumulate until

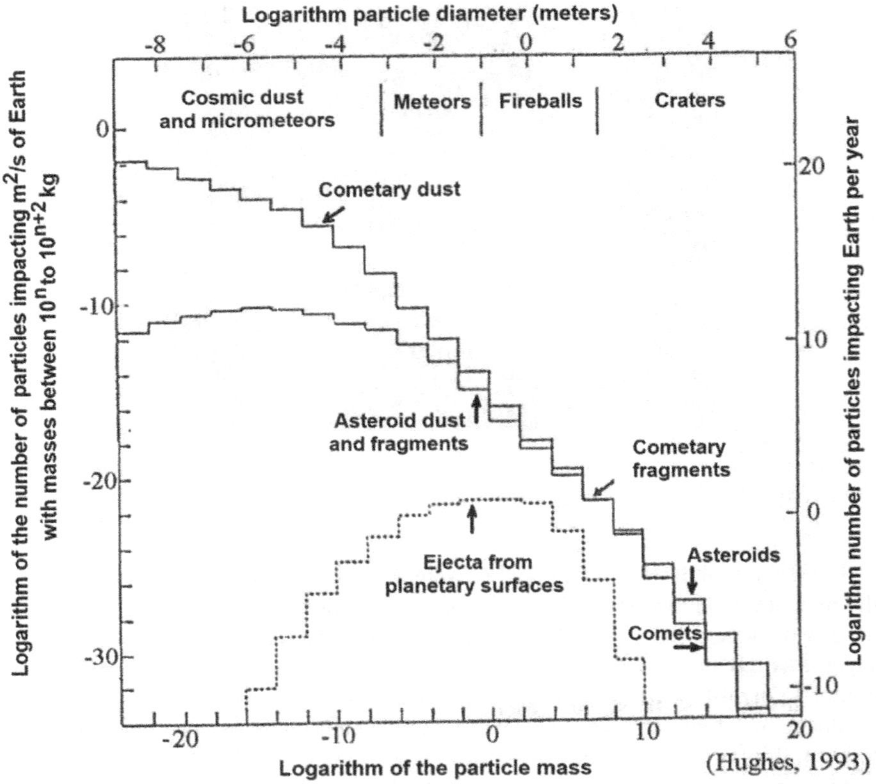

Fig. 1.5 Flux of materials reaching the Earth's surface annually. The logarithm of their mass at atmospheric entry is plotted against the number (figure adapted from Hughes, 1993)

they are found. Deserts such as the Sahara are the source of a huge number of the meteorites that exist in specially preserved collections in museums and research centres around the world. Certain regions of the globe, due to their extreme dryness, seem to behave as meteorite storehouses. One example is the north-west region of Africa, where meteorites are classified under the acronym NWA (North West Africa), of which about 6,700 have been catalogued up to June 2012. These are actually a small part of the tens of thousands found by Bedouins, travellers and adventurers but, unfortunately, many of them are sold to the highest bidder and have no record of the exact location where they were found. Such a process makes it difficult for most NWA meteorites to be classified although programmes can be organised for other specimens perhaps associated with the same fall. Some of the many meteorites that tourists buy from traders in certain African countries end their days forgotten in the drawers of collectors and, if not later donated or loaned for study, are inexorably lost to science.

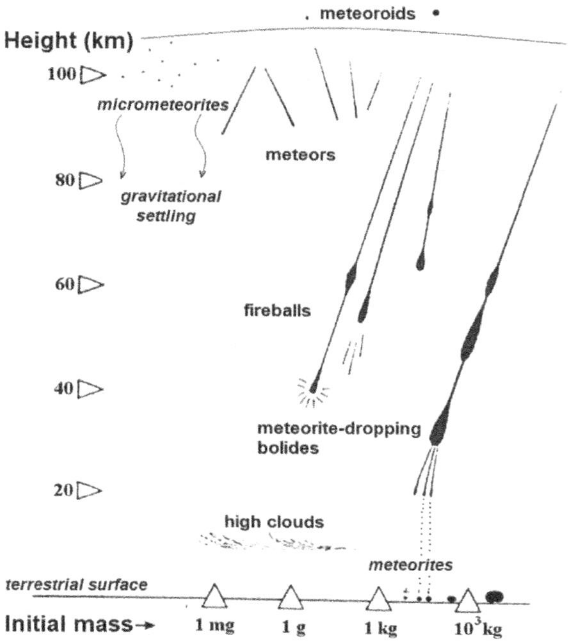

Fig. 1.6 Meteoric phenomena at various scales (adapted from J. Rendtel et al. (1995) with permission)

This is not always the case, however, and the wider dissemination of the teachings of this science means that meteorite discoverers collaborate with us. A valuable example was the recovery of the Ardón meteorite in León in 1931. On 9 July 1931, at 10.30 a.m., a huge fireball flew over the province of León and generated a series of audible bursts from the capital and other nearby municipalities, including Boñar and Cistierna, as reported in the Spanish media at the time. Rosa González Pérez, then a girl of 11 years old, was on the outskirts of the municipality of Ardón when she heard a roar that came from a trail of dust. Right in front of her she saw a small smoking rock fall from the sky and, when she picked it up, she noticed that it was still hot. Out of ignorance, she said nothing about her find and kept it in a small box, preserving it in very good condition for more than 80 years. It was a nephew of this lady, José Antonio González, who years later thought that this small blackish rock of 5.5 g could be important. When he contacted the author of these lines and sent the specimen, it was quickly identified as a meteorite (Trigo-Rodríguez et al., 2014).

The Ardón fall may not be an isolated case, as the number of recorded meteorite falls in Spain is much lower than statistics would suggest. In fact, our recovery of the Villalbeto de la Peña meteorite in 2004 occurred more than 56 years after the Reliegos fall (1947). This too-long interval coincided with the dictatorship, so we suspect that some meteorites may have been hidden or sold into private collections. For this reason, the 2007 Law on Natural Heritage and Biodiversity (Law 42/2007) now recognises that meteorites in Spain are geological heritage and, therefore, must be preserved and remain in the country.

On the other hand, the polar regions, especially Antarctica, are particularly favourable for the recovery of meteorites and remain outside the meteorite trade thanks to the efforts of several international organisations. There are various processes by which meteorites accumulate and can be recovered even after they have been part of huge masses of glacial ice (Harvey, 2003). In fact, scientific interest in promoting comprehensive searches in these regions arose after the designation of the International Year of Geophysics in 1957. Between 1961 and 1964 a number of expeditions carried out in that context recovered several metallic meteorites, some of which appeared to be associated with glacial moraines. However, one remarkable event was to mark a major breakthrough in our understanding of the preservation of meteorites in polar regions and the importance of their study. In late 1969, a group of Japanese scientists discovered nine meteorites of five different petrographic types in a relatively restricted area of the Yamato Mountains which they were exploring as part of the tenth Japanese scientific expedition to Antarctica (JARE-10). The accumulation of meteorites was associated with the movement and abrasion of ice in the region that had concentrated the specimens to naturally facilitate their discovery together (Yoshida et al., 1971).

William Cassidy of the University of Pittsburgh took it upon himself to promote in the United States a systematic search similar to the successful Japanese programme. The *Antarctic Search for Meteorites (ANSMET)* programme was born in 1975. The Spanish geologist Francisco Anguita was an exceptional participant in the expedition carried out in 1991–92, the most successful up to then, with 220 kg of meteorites recovered (Anguita and Castilla, 2003). So far, many international research groups have discovered nearly 40,000 meteorites in the region. To give an idea of their importance, Antarctic meteorites account for about 70% of all those found in terrestrial collections.

It is worth using this book to say that those of us who are professionally engaged in the study of meteorites want them to be recognised as part of our common heritage. They should remain in collections for public display but also to a large extent be accessible for study by researchers with well-reasoned projects and who undertake to return them to the conservation centres within the time required to complete the work. This is the procedure of major museums around the world and also of NASA itself. Meteorites provide a scientific opportunity, often unrepeatable, to address key questions about the physico-chemical processes that have occurred since the very origins of the solar system in the bodies from which they originate. For example, by studying specimens from NASA's Antarctic collection, we discovered a tiny fragment of a primordial comet whose chemistry predates the materials we know, and we have found meteorites whose reflective properties resemble those of certain peculiar asteroids (Moyano-Cambero et al., 2016) or comets (Tanbakouei et al., 2020).

The meteorites recovered in Spain, thanks to the efforts of the members of the *Red de Investigación sobre Bólidos y Meteoritos (SPMN;* www.spmn.uji.es*)* and their collaborators, are available to museums and research centres that act as *hosting institutions for the Meteoritical Society*, so that other researchers from around the world can study them upon request. Meteorites were not legislated for in Spain until 2007, the year in which we managed to have both fallen meteorites and those

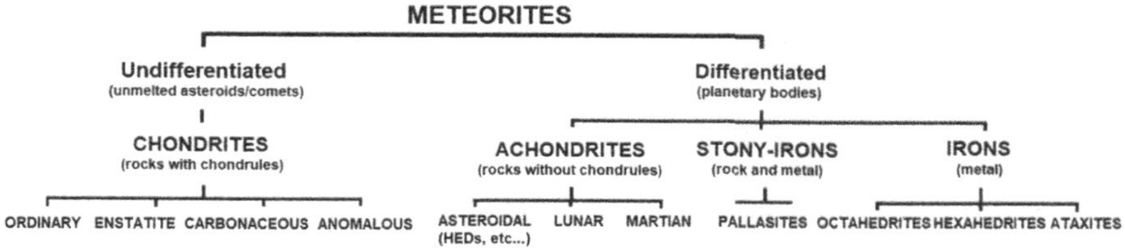

Fig. 1.7 Diagram of the main classes of meteorites (Trigo-Rodríguez, 2012a)

found in Spanish territory included as geological heritage, and therefore their future trade was prohibited (Law 45–2007 on Spanish Geological Heritage). Obviously, in every country there are different regulations on this matter, but the important thing is preserving these fascinating rocks.

A Brief Introduction to The Types of Meteorites

From the very birth of meteoritics, a simple classification of meteorites reaching the Earth was made from observing the main nature of these rocks: rocky, metallic and rocky-metal. Today, the established groups are based on chemical and mineralogical criteria (Fig. 1.7). For the stony and metallic meteorites, there were other names that are nowadays obsolete, such as aerolites or siderolites.

A more coherent classification is to identify them according to their degree of processing in the parent body. We call *undifferentiated* meteorites all those that come from bodies a few hundred kilometres in diameter that have therefore not undergone chemical differentiation. In general, such meteorites are called *chondrites*. This name comes from the fact that in their structure we find mostly an amalgam of glassy spherules of igneous nature called *chondrules*, together with other types of particles of different origin. All these aggregate components in the chondrites were part of the so-called *protoplanetary disc* long before the planets or the minor bodies of the solar system were formed. About 85% of the meteorites observed to fall are chondrites, and of these about 90% are compacted rocks formed by a process called *metamorphism*. This is basically a change of minerals or geologic texture in primordial existing chondrites (known as petrologic type 3 chondrites) that occurs due to heat and pressure—so it might be produced by a compression suffered in depth in the parent asteroid, or by a shock process induced by a collision. Increasing heat and pressure brings chondrites to petrological types 4 to 6, and their rock-forming materials are compacted accordingly.

Obviously, not all the rocks reaching the Earth are chondrites. Many bodies reached sizes over 1,000 km and experienced *differentiation*. Their primordial minerals were melted, and formed igneous mixtures of other secondary minerals. For that reason, in the right-hand side of the diagram in Fig. 1.7 we find meteorites originating from differentiated planetary bodies. We can distinguish three main types: rocky (or achondrites), stony-irons and irons. Their components are the result of igneous processes occurring in bodies thousands of kilometres in diameter. Although they may retain chemical and isotopic signatures of the primordial minerals, their materials consist of secondary minerals, created in their parent planetary bodies. In fact, the most important sources of achondrites are the Moon, Mars, and Vesta. Different lines of evidence, such as the existence of achondrites that do not seem to be associated with the above-mentioned groups, indicate that achondrites must exist as remnants of planetary bodies destroyed in the past or that we have not yet characterised. This is not surprising given the absence of sample return missions that allow us to delineate the peculiar chemical signatures of distant objects.

Other meteorites come from regions of the inner mantle of differentiated bodies that must have been excavated in large impacts. This is the case with stony-iron and metallic meteorites. The most common stony-iron meteorites are the *pallasites* which, when cut and polished, reveal themselves to be a precious mixture of olivine crystals that appear to be set in a strong iron-nickel alloy matrix.

Finally, metallic meteorites arrive on Earth from the cores of planetary bodies destroyed in large impacts. These meteorites are thus the densest materials we have access to and can be considered representative of those that form the Earth's core.

1.4 The Value of Recovering Meteorites

Thus, meteorites are samples from other worlds that, similar to terrestrial rocks, can impart immeasurable information about their origin and formation times. Meteorites are not a panacea, because of the millions of bodies in the solar system, we receive meteorites from about 15 undifferentiated and about 80 differentiated asteroids (Eugster et al., 2006). There are several reasons to explain this. It is important to remark that there is a significant bias concerning the survival of meteoroids that reach the ground as meteorites. High-strength meteoroids are the most common survivors, so about 90% of our meteorite collections are dominated by dense rocks such as highly metamorphosed chondrites, stony-irons, irons or differentiated meteorites of the style of basalts.

On the other hand, there are limitations to the number of objects sampled by nature imposed by the fact that we only receive fragments of bodies that have fallen into certain unstable zones (or resonances) where they are redirected towards Earth. The transport mechanisms of these rocks from their parent bodies to Earth are not nearly as fast as we might think. On average, since they are detached after colossal impacts from the surface of their parent bodies, they usually spend several tens of millions of years in interplanetary space. We know this because noble gases are implanted on the surface of these rocks, which are constantly emitted from the Sun as part of the solar wind. A series of cosmic coincidences will allow these rocks to find our planet on their way, and only in the most favourable geometrical conditions will the velocity and angle of incidence be adequate to allow the survival of debris that will reach the Earth's surface in the form of meteorites.

When they reach Earth, they do so by attracting our attention. Typically, at these enormous speeds more than 97% of the mass of these rocks will be lost in the fireball phase, producing ionised gas, fine dust and a small percentage of light, but enough to make these phenomena extraordinarily striking to the lucky casual observer. However, most meteorites reach the Earth over remote and uninhabited areas and are rarely observed in their fall.

The first photographic fireball recording programmes were initiated in the Czech Republic by the astronomer Zdenek Ceplecha (1929–2009) of the Ondřejov Observatory. This pioneer started with photographs taken of a large fireball from which the Príbram meteorite was recovered in 1959 (Ceplecha, 1961). From the images obtained from several stations, the orbit of the parent meteoroid in the solar system could be obtained. This unprecedented success prompted the birth of two dedicated networks, one in the Czech Republic led by Ceplecha and the other in the United States led by Richard E. McCrosky (1924–2012) (McCrosky and Ceplecha, 1968). In fact, McCrosky also involved the prestigious Smithsonian Astrophysical Observatory to create the *Prairie Network*, which consisted of 16 photographic stations that monitored the sky to identify large meteor-producing bolides around the state of Nebraska. On 3 January 1970 a large fireball was recorded over Oklahoma and reconstruction of its trajectory allowed the recovery of the Lost City meteorite (McCrosky et al., 1971). A total of four fragments were recovered, totalling 17 kg of ordinary H5 class chondrites, but funding for the project came to an end some years later. It was a Canadian photographic network led by I. Halliday who recovered the Innisfree meteorite in 1977 (Halliday et al., 1978).

Today, meteorite recovery has become a field of opportunity in the study of planetary sciences. The calculation of their orbits and, based on these, of the dynamic transport mechanisms towards the Earth and the association with their original bodies are some of the main scientific objectives of the SPMN research network on bolides and meteorites (www.spmn.uji.es), which was created almost 20 years ago by several researchers interested in these phenomena. Thanks to our joint efforts, we were able to recover the Villalbeto de la Peña ordinary chondrite that fell in the municipality of the same name in Palencia on 4 January 2004 (Llorca et al., 2005), from which we obtained the ninth orbit of a meteorite (Trigo-Rodríguez et al., 2006) and the Puerto Lápice eucrite (Llorca et al., 2009; Trigo-Rodríguez et al., 2009). As an added value to these studies, we have managed to relegate the UFO phenomenon and give a rational explanation to the fireballs that cross Iberian Peninsula skies.

In general, the photographic programmes for fireball detection that we used in the 1980s have become obsolete and are giving way to programmes for continuous monitoring of the firmament based on CCD cameras and video detection in which the

SPMN network has been a pioneer (Trigo-Rodríguez et al., 2004; Madiedo and Trigo-Rodríguez, 2008). In this sense, the models of meteoritic atmospheric trajectories obtained and the study of the images in sequential mode allow a better understanding of the fragmentation processes of meteorites arriving on Earth.

In addition to meteorites falling to Earth, and Moon rocks brought back by the astronauts of NASA's Apollo missions, particles recovered by the Stardust probe when passing through the coma of Comet 81P/Wild 2 (Brownlee et al., 2006), regolith grains from the asteroid Itokawa captured by Hayabusa probe (Nakamura et al., 2012) and our samples obtained by Hayabusa 2 from asteroid 162,173 Ryugu have recently been added to our collection of extraterrestrial materials. Soon the OSIRIS-REx mission will bring samples from primitive asteroid 101,955 Bennu. The lessons learned about these materials and the first sample return missions that have obtained them will be discussed in another chapter.

Questions for Chapter 1

Q.1. As previously explained, the global flux of extraterrestrial material on Earth is dominated by sub-millimetre particles that reach the top of the atmosphere as fragile aggregates, but others reach as bigger centimetre-sized meteoroids that in the appropriated geometric circumstances reach the ground as meteorites or micrometeorites. In total, the mass arrived by this bimodal flux is about 40 million kg, while the sub-millimetre component has recently been estimated to be 15 million kg (Rojas et al., 2021). a) Calculate the approximate mass in grams of extraterrestrial material reaching a square meter of the Earth, and b) Assuming that an individual micrometeorite weights about 1×10^{-5} g, how many of these particles are reaching each square metre per year? (Note: Consider that the surface of the Earth is 5.1×10^{14} m^2) [Solution: a) Taking into account that the Earth's total surface area is 5.1×10^8 km^2, every square metre receives about 0.1 mg annually. b) If the flux was typically made by these small particles, every year we will get about ten per square metre.]

Q.2. The total flux of meteorites larger than 1 kg was estimated by studying the number of meteorite-dropping bolides by Halliday et al. (1978) using the Canadian camera network. They computed that a meteorite fall with at least 1 kg of meteorites occurs in an area of 1 million km^2. (a) How many meteorites are expected in the United States per year? and (b) Compare the number of expected meteorite falls in the US, or in your own country, with the number of meteorite falls actually recovered lately. This comparison will exemplify that finding a meteorite is a tricky task! (Note: You can use the online Meteoritical Bulletin Database, given in online resources, to know the number of meteorite falls recovered, or the meteorites recovered in your country.) [Solution: (a) US total surface area is 9,831,510 square km, so about ten meteorite falls occur every single year. Obviously, not so many are recovered as those occurring in forests or remote areas are unnoticed].

Q.3. To recover meteorites after their falls is extremely difficult. Numbers are often biased towards high-strength meteorites that better survive the deceleration in the atmosphere. We can use the Meteoritical Bulletin Database to exemplify this. The following table allows us to infer the percentage of meteorites of each main class being recovered after their falls. While the carbonaceous chondrite flux seems to be dominant by studying the material implanted in the Lunar regolith, the fact is that the number of meteorites of this class is relatively minor. (a) Why do you think that the relative number of recovered carbonaceous chondrites is minor? (b) Using similar arguments, why do you think that the ordinary chondrites are dominant? (c) If the iron-rich meteorites are expected to be denser than stony meteorites, why do you think their fall numbers are so low?

[Solution: (a) There are at least two important factors for the lack of carbonaceous chondrite falls. First, carbonaceous chondrites are porous meteorites that exhibit the lowest bulk densities known (Britt and Consolmagno, 2003). In consequence, they are fragile and their survival in space, under radiation and impacts, is limited, as also demonstrated by the study of their Cosmic Ray Exposure Ages (see Fig. 5.7 and discussion in Chap. 5). Second, as they are composed of friable, fine-grained materials, they cannot survive atmospheric interactions as large stones, so they are far more difficult to find. For example, Revelstoke, the first carbonaceous chondrite recovered after recording its bright fireball in 1965, weighs only 1 g, and it could only have been recovered because it was distinguishable over snow. Most of the original meteoroid that produced that fall arrived to the ground as dust. (b) The ordinary chondrites exhibit higher strength and density than the carbonaceous ones. In addition, there are several asteroid families in the main asteroid belt that are formed by S-class asteroids from which the ordinary chondrites are released. (c) Falls from irons and stony irons are scarce. Iron-rich meteorites in between the population of bodies approaching the Earth and forming the main asteroid belt are also rare. The reader is encouraged to expand on these answers and look for additional information.]

Meteorite class	Number of meteorite falls	Percentage
Ordinary chondrites	948	70
Carbonaceous chondrites	51	3.7
Achondrites	86	6.3
Stony-irons (pallasites and mesosiderites)	10	0.7
Irons	47	3.5
Other classes/doubtful	214	15.8
All	1356	100

Online Resources

Meteorites at the American Museum of Natural History: https://www.amnh.org/exhibitions/permanent/meteorites

Catalogue of Meteorites of the National History Museum: https://www.nhm.ac.uk/our-science/data/metcat/

Meteor Data Center: http://www.astro.amu.edu.pl/~jopek/MDC2007/

Meteoritical Bulletin Database: https://www.lpi.usra.edu/meteor/metbull.php

Spanish Bolide and Meteorite Research Network: www.spmn.uji.es

References

Anguita and Castilla, 2003.Anguita F. and G. Castilla (2003) *Crónicas del Sistema Solar*, Equipo Sirius, Madrid.

Borovicka J. (1993) A fireball spectrum analysis. *Astronomy and Astrophysics* 279, 627–645.

Britt D.T. and G.J. Consolmagno (2003) Stony meteorite porosities and densities: A review of the data through 2001. *Meteoritics & Planetary Science* 38, 1161–1180.

Brownlee et al., 2006.Brownlee D. et al. (2006) Comet 81P/Wild 2 under a microscope. *Science* 314, 1711–1716.

Brownlee D.E. (2001) The origin and properties of dust impacting the Earth Accretion of Extraterrestrial Matter Throughout Earth's History, eds. B. Peucker-Ehrenbrink and B Schmitz (New York: Kluwer Academic/Plenum) pp 1–12

Ceplecha Z. (1961) Multiple fall of Pribam meteorites photographed. *Bulletin of the Astronomical Institutes of Czechoslovakia* 12, 21–47.

Eugster O. et al. (2006) Irradiation records, cosmic-ray exposure ages, and transfer times of meteorites. In *Meteorites and the Early Solar System II*, D. Lauretta and H.Y. McSween (eds.), The University of Arizona Press, Tucson, AZ, USA, pp. 829–851.

Halliday I., A.T. Blackwell and A.A. Griffin (1978) The Innisfree meteorite and the Canadian camera network. *Journal of the Royal Astronomical Society of Canada* 72, 15–39.

Harvey, R.P. (2003) The origin and significance of Antarctic meteorites. *Chemie der Erde – Geochemistry* 63, 93–147.

Hörz et al., 2006.Hörz F. et al. [including J.M. Trigo-Rodríguez] (2006) Stardust implications for Wild 2 coma dust. *Science* 314, 1716–1719.

Hughes D.W. (1993) Meteoroids: an overview. In *Meteoroids and their Parent Bodies, Proceedings of the International Astronomical Symposium held at Smolenice, Slovakia, July 6–12, 1992, Bratislava: Astronomical Institute, Slovak Academy of Sciences, 1993*, J. Stohl and I.P. Williams (eds.), 15–28.

Jenniskens, 2006.Jenniskens P. (2006) *Meteor showers and their Parent Comets*. Cambridge University Press, Cambridge, UK, 790 pp.

Llorca J., J.M. Trigo-Rodríguez, J.L. Ortiz, J.A. Docobo, J. Garcia-Guinea, A.J. Castro-Tirado, A.E. Rubin, O. Eugster, W. Edwards, M. Laubenstein and I. Casanova (2005) The Villalbeto de la Peña meteorite fall: I. Fireball energy, meteorite recovery, strewn field and petrography. *Meteoritics & Planetary Science* 40, 795–804.

Llorca J., I. Casanova, J. M. Trigo-Rodríguez, J. M. Madiedo, J. Roszjar, A. Bischoff, U. Ott, I. A. Franchi, R. C. Greenwood and M. Laubenstein (2009) The Puerto Lápice eucrite. *Meteoritics and Planetary Science* 44, 159–174.

Madiedo and Trigo-Rodríguez, 2008.Madiedo J.M. and J.M. Trigo-Rodríguez (2008) Multi-station video orbits of minor meteor showers. *Earth Moon and Planets* 102, 133–139.

McCrosky R.E. and Z. Ceplecha (1968) Photographic networks for fireballs. *SAO Special Report No 288*, 19 pp.

McCrosky R.E., A. Posen, G. Schwartz and C.-Y. Shao (1971) Lost city meteorite: its recovery and a comparison with other fireballs. *Journal of Geophysical Research* 76, 4090–4108.

McSween, 1999.McSween H.Y. (1999) *Meteorites and their Parent Bodies*. Cambridge University Press, Cambridge, UK.

Moyano-Cambero C.E., J.M. Trigo-Rodríguez, J. Llorca, S. Fornasier, M.A. Barucci and A. Rimola (2016) A plausible link between the asteroid 21 Lutetia and CH carbonaceous chondrites. *Meteoritics & Planetary Science* 51, 1795–1812.

Nakamura E. et al. (2012) Space environment of an asteroid preserved on micrograins returned by the Hayabusa spacecraft. *Proceedings of the National Academy of Sciences*, doi: pnas.1116236109.

Rendtel et al., 1995.Rendtel J., R. Arlt and A. McBeath (1995) *Handbook for Visual Meteor Observers*. International Meteor Organization Publication, Antwerp, Belgium.

Rojas J., J. Duprat, C. Engrand, E. Dartois, L. Delauche, M. Godard, M. Gounell, J.D. Carrillo-Sánchez, P. Pokorný and J.M.C. Plane (2021) The micrometeorite flux at Dome C (Antarctica), monitoring the accretion of extraterrestrial dust on Earth. *Earth and Planetary Science Letters* 560, article 116794, 11 pp.

Tanbakouei S., J.M. Trigo-Rodríguez, J. Blum, I. Williams and J. Llorca (2020) Comparing the reflectivity of ungrouped carbonaceous chondrites with that of short period comets like 2P/Encke. *Astronomy & Astrophysics* 641, A58, 7 pp.

Trigo-Rodríguez J.M. (2012a) *Meteoritos, Colección ¿Qué sabemos de...*, Editorial Catarata-CSIC, Madrid.

Trigo-Rodríguez, 2012b.Trigo-Rodríguez J.M. (2012b) *Las raíces cósmicas de la vida, Colección El Espejo y la Lámpara*. Servei Publicacions UAB, Barcelona.

Trigo-Rodríguez J.M. (2019) The flux of meteoroids over time: meteor emission spectroscopy and the delivery of volatiles and chondritic materials to Earth. In *Hypersonic Meteoroid Entry Physics*, G. Colonna, M. Capitelli and A. Laricchiuta (eds.), Institute of Physics Publishing, IOP Series in Plasma Physics, pp. 4–1/4–23. Online ISBN: 978–0–7503–1668–2.

Trigo-Rodríguez J.M., and Blum J. (2021) Learning about comets from the study of mass distributions and fluxes of meteoroid streams. *Monthly Notices of the Royal Astronomical Society*, stab 2827, 13 pp.

Trigo-Rodríguez J.M., A. Castro-Tirado, J. Llorca, J. Fabregat, V. J. Martínez, V. Reglero, M. Jelínek, P. Kubánek, T. Mateo and A. de Ugarte Postigo (2004) The development of the Spanish Fireball Network using a new all-sky CCD system. *Earth, Moon Planets* 95, 553–567.

Trigo-Rodríguez J.M., J. Borovička, P. Spurný, J.L. Ortiz, J.A. Docobo, A.J. Castro-Tirado and J. Llorca (2006) The Villalbeto de la Peña meteorite fall: II. Determination of the atmospheric trajectory and orbit. *Meteoritics & Planetary Science* 41, 505–517.

Trigo-Rodríguez J.M., J. Borovička, J. Llorca, J. M. Madiedo, J. Zamorano, and J. Izquierdo (2009) Puerto Lápice eucrite fall: strewn field, physical description, probable fireball trajectory, and orbit. *Meteoritics and Planetary Science* 44, 175–186.

Trigo-Rodríguez J.M. et al. (2013) The 2011 October Draconids outburst. I. Orbital elements, meteoroid fluxes and 21P/Giacobini-Zinner delivered mass to Earth, *Monthly Notices of the Royal Astronomical Society* 433, 560–570.

Trigo-Rodríguez J.M., J. Llorca, M. Weyrauch, A. Bischoff, C.E. Moyano-Cambero, K. Keil, M. Laubenstein, A. Pack, J.M. Madiedo, J. Alonso-Azcárate, M. Riebe, R. Wieler, U. Ott, M. Tapia and N. Mestres (2014) The Ardón L6 ordinary chondrite: A long hidden Spanish meteorite fall, *Meteoritics & Planetary Science* 49(8), 1475–1484.

Yoshida et al., 1971.Yoshida M., H. Ando, K. Omoto, R. Naruse, Y. Ageta (1971) Discovery of meteorites near Yamato Mountains, East Antarctica. *Antarctic Rec.*, 39, pp. 62-65

Chapter 2
Meteorite Parent Bodies and Their Routes to Earth

It is evident that asteroids are the main bodies that produce meteorites. While we have doubts about the capacity of comets to do so, these have been disappearing in recent decades, at least when talking about evolved comets (Trigo-Rodríguez, 2015; Trigo-Rodríguez and Williams, 2017). Even so, the Earth also receives rocks from various planetary bodies such as the Moon, Vesta and Mars. This chapter aims to provide insight into the various regions of the solar system populated by these minor bodies and to explain the dynamic mechanisms that allow the transport of meteorites to Earth. What we learn about such physical processes is important because they also act to deflect asteroids from the main belt into the near-Earth region.

An important source of information are the orbits of large meteoroids, particularly those with a diameter of between one and ten metres, which manage to survive their abrupt passage through the atmosphere to produce meteorites. We will see how orbits reconstructed from the fireball record allow us to identify certain areas of the main belt as the source of these rocks. At the same time, when meteorites are recovered, they provide very important clues because their interiors have been exposed to radiation from interplanetary space. In that sense, isotopic criteria have been developed to date the transport time from when the meteorite-producing rocks were on their asteroids until they reached the Earth. In particular, this is achieved by determining *Cosmic-ray exposure ages* or *CREAs* (see e.g., Eugster et al., 2006). These cosmic-ray exposure periods are variable but show that many of these rocks have been generated as a result of the same collision between asteroids, revealing that our planetary system is dynamic and transport is subject to collisions in the main asteroid belt on scales of millions of years (Rubin, 2002). From the CREAs we know that most of the meteorites that reach Earth were detached tens of millions of years ago from their parent asteroids as a consequence of colossal impacts of which most of today's asteroids are but the remnants (Fig. 2.1). In order not to complicate these introductory chapters, in Chap. 5 I will explain in more detail the application of CREAs to find out the residence times of meteorites in interplanetary space.

J. M. Trigo-Rodríguez, *Asteroid Impact Risk*, Impact Studies,
https://doi.org/10.1007/978-3-030-95124-5_2

Fig. 2.1 The asteroids Mathilde, Gaspra and Ida to scale. Mathilde appears brighter in this image than it actually is, as its formative materials are dark and closely related with the carbonaceous chondrites (NASA/NEAR Shoemaker/Galileo)

2.1 The Origin of Asteroids and Meteorites

In the scientific literature, the development of the first theories of accretion of small rocky blocks leading to the final formation of planets led to the coining of the term *planetesimal*. This term refers to the first solid blocks, with diameters ranging from metres to a few kilometres, that emerged as a result of the aggregation of materials formed in the protoplanetary disc some 4.56 billion years ago. Planetesimals grew from collisions between highly porous aggregates formed by the first solid materials that condensed in the protoplanetary disc formed around the Sun. We know this because the most primitive meteorites, the so-called carbonaceous chondrites, never melted but were compacted by impacts and retained the primordial components (Wasson, 1985; Nittler et al., 2019). These components can be dated to better understand the origin of the materials that form the undifferentiated asteroids.

Thus, the asteroids we find today in different regions of the solar system are the remnants of those early bodies, shaped by countless impacts and dynamical processes over the eons (Fig. 2.1). This chapter describes the origin and location of the surviving bodies today. As described below, depending on the distance to the Sun, objects of different nature formed and gave rise to the enormous diversity of chondritic asteroids. In the outer regions of the disc, amalgams of small rocky particles, ices and organic matter would form and grow to become the first cometesimals. Such objects today would be represented by comets, although comets may also have undergone major modification processes as a result of collisions with other objects. In the inner region, where the terrestrial-type planets are found, the largest blocks would be rocky in nature and would tend to form asteroids.

Most of the meteorites reaching Earth come from asteroids and, to a lesser extent, we suspect also from certain comets (Trigo-Rodríguez, 2015). Even so, a small percentage of differentiated meteorites (known as *achondrites*) reach our planet from the Moon, Mars or asteroid 4 Vesta (Grady, 2000). This is demonstrated by their chemical and mineralogical properties, but also by statistics collected over the last century for the main classes. However, it is extremely difficult to establish a conclusive origin for most meteorites. In other words, we know that most meteorites come from asteroids, but to establish from which ones is a challenge that has only been achieved in a few cases. On the other hand, meteorites contain chemical elements that are or were radioactive and whose study allows the establishment of isotopic chronometers to date the formation processes of the asteroids from which they originate.

2.2 The Main Asteroid Belt

The main asteroid store is the so-called main belt. In this region between the orbits of Mars and Jupiter, more than 700,000 asteroids are known today (Fig. 2.2), but this number is growing all the time. In fact, we know that the region is home to many more, because at the limit of the asteroid definition—i.e. small objects with a diameter of at least 10 m—the number would rise to several million. In fact, the asteroid belt hosts a majority of relatively small objects with diameters of less than 300 km. Table 2.1 shows the largest ones.

With such overwhelming numbers showing that asteroids are very numerous, it is curious to think that just over two centuries ago we did not know of any asteroids at all. The scientific revolution that led to their discovery began in the search for

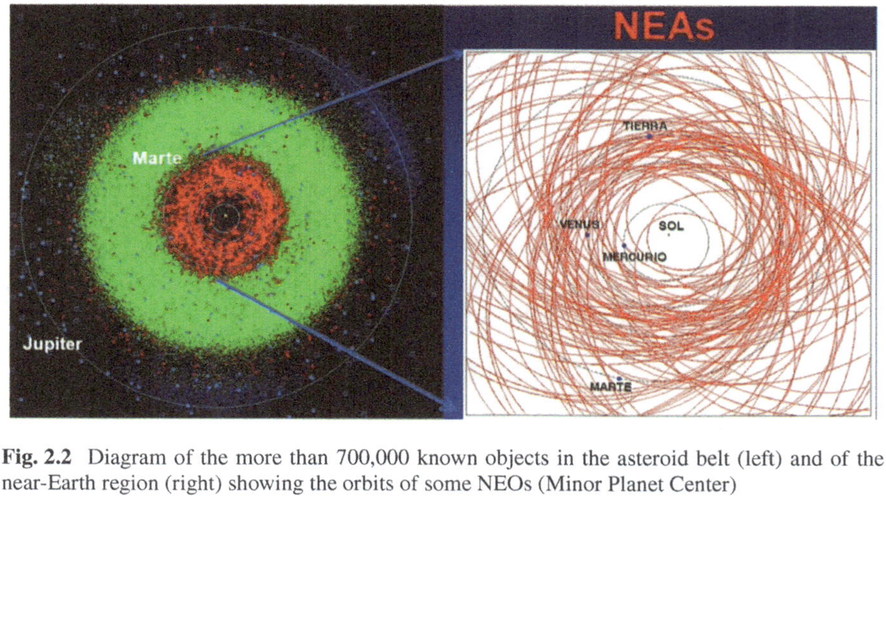

Fig. 2.2 Diagram of the more than 700,000 known objects in the asteroid belt (left) and of the near-Earth region (right) showing the orbits of some NEOs (Minor Planet Center)

Table 2.1 Asteroids with a diameter greater than 300 km, indicating their diameter, mass and mean density (where known)

Asteroid	Diameter (km)	Mass (kg)	Average density (g/cm^3)	Spectral class
1 Ceres	913	9.5×10^{20}	2.1	G
2 Pallas	523	2.4×10^{20}	2.9	B
4 Vesta	520	2.7×10^{20}	3.5	V
10 Hygiea	429	1.1×10^{20}	-	C
511 Davida	337	1.1×10^{20}	-	C
704 Interamnia	333	7.0×10^{19}	-	F
52 Europe	312	5.2×10^{19}	-	C

other planets that was spurred by Johannes Kepler (1571–1630), who first suggested the existence of a planet between Mars and Jupiter in his work *Mysterium Cosmographicum* published in 1596. He proposed this when he realised that Mars and Jupiter were too far apart compared to other planets. Taking into account that the average distance from the Earth to the Sun is defined as 1 astronomical unit (AU, equivalent to about 150 million km), Mars is on average 1.524 AU away while the giant Jupiter is 5,203 AU away. It also seemed too great a distance to Isaac Newton (1643–1727) and Immanuel Kant (1724–1804). In Newton's publication of *Principia Mathematica* in 1687, he introduced the idea that a single force, gravity, was responsible to explain the movements of the planets, and other minor bodies such as comets. The success in explaining the movement of planets motivated many astronomers to study the movements with increasing astrometric accuracy, adding new tests to the theory. In addition, the different parameters describing the heliocentric movements of the planets were carefully studied, so different hypotheses about the formation of the planetary system appeared at that time. Some astronomers noticed that the average distance of the planets to the Sun was not random, and developed some novel approaches (see The Titius–Bode Law).

The Titius–Bode 'Law

The distribution of the planets surrounding the Sun seems to be not random, but following a pattern. The first elegant explanation came from Johann Daniel Titius (1729–1796) and Johann Elert Bode (1747–1826) who independently established a mathematical progression known as the Titius–Bode Law. The equation makes it possible to predict the average distance to the Sun (D, in astronomical units AU) of the known planets:

$$D = 0.4 + 0.3 \times (2)^x$$

where on x we should state: $-\infty$, 0, 1, 1, 2, 3....

Obviously, this is just a mnemonic rule in which the mean distance to the Sun (D) appears to follow a regular pattern. It seems that there is not an obvious physical foundation, but certainly at the early time of protoplanetary disc formation the distribution of planetesimals around the Sun was not uniform.

Astronomers did not agree on the physical origin of this numerical relationship, but it seemed to well define the order occupied by the planets around the Sun. However, there was an empty space between Mars and Jupiter so it served to start a debate. In addition, in 1781 the astronomer Friedrich Wilhelm Herschel (1738–1822) discovered Uranus at a distance similar to that predicted by the mnemonic rule. The discovery promoted a growing interest among European astronomers in discovering new planets, and the mnemonic rule became really popular and was much used for predictions.

As a consequence of another prediction of the Titius–Bode relationship, an important group of German astronomers led by Franz X. von Zach (1754–1832), director of the Seeberg Observatory, developed a strategy to search for the planet that should exist (according to that rule) between the orbits of planets Mars and Jupiter. The undertaking was enormous, as those involved intended to distribute their work monitoring the zodiacal band, the region where the planets move, in order to develop a joint programme capable of finding the missing planet. On 20 September 1800, von Zach met with five other German astronomers, including Bode, to divide this band into 24 regions in which to meticulously scrutinise the possible movement of faint stars. At the meeting they selected other astronomers operating prestigious observatories who would be able to support them in the mammoth endeavour.

Among the astronomers selected by von Zach's team was the Italian Giuseppe Piazzi (1746–1826), who on 1 January 1801 discovered Ceres without even being aware that he had been commissioned to search for it. But the discovery of this dwarf planet (Fig. 2.3) was not by chance; the Italian was working on a stellar catalogue in which he accurately measured the position of the fixed stars. On the first of January he noticed that one of these stars had been showing a movement of its own for several nights. Although at first he thought he had made a measurement error, by the fourth of January he was convinced that he had discovered a new object which, a priori, he thought was a comet. In fact, he announced it to the press as such, although within two weeks he suspected that it might be a new planet, as he wrote to his friend Barnaba Oriani (1752–1832). However, he was rather more reticent to suggest such a discovery in a letter to Bode, perhaps hoping for independent confirmation.

After a few weeks the news of the discovery had passed from newspaper to newspaper and spread throughout the European continent. It was then that the Frenchman Joseph Lalande (1732–1807) asked Piazzi for the derived coordinates of the new object in order to obtain its orbit. He immediately realised that it conformed perfectly to a circular orbit, as expected for a new planet, and that it also conformed to the distance suggested by the Titius–Bode Law. He then alerted von Zach and spread

Fig. 2.3 Comparison between the diameters of Mars, Ceres and Vesta (NASA/JPL-Caltech/UCLA/MPS/DLR/IDA)

the word among the Prussian academy of sciences, including Bode. It was the latter who decided to propose to the academy that the star should be called Hera or Juno, without taking into account the wishes of Piazzi, its discoverer. The Italian's anger was monumental and he decided to insist in all forums that the star should be called Ceres Ferdinandea, a complex name in honour of the patron goddess of Sicily and King Ferdinand I of the Two Sicilies. In the end, the astronomical community decided to accept the short version of the proposal and so it is the asteroid known as 1 Ceres, visited in 2015 by the Dawn mission (NASA). The number in front of the name indicates the asteroid's catalogue number as defined by the Minor Planet Center (MPC).

In this historical review of the discovery of Ceres, one important fact remains to be mentioned. When Piazzi discovered Ceres, this dwarf planet was about to enter conjunction with the Sun. This was equivalent to losing it for a few months in its own unobservable close-to-the-Sun motion. Everyone feared that it would be lost forever, because in practice only an arc of about 3° had been observed in its apparent trajectory across the celestial vault. Fortunately, the renowned mathematician Carl Friedrich Gauss (1777–1855) developed a very precise mathematical method for fitting a perfect ellipse to a small arc of observation of a star. These very important theoretical advances made it possible to determine precisely the orbital elements defining the orbit of the star and to predict the celestial coordinates at which it should be sought after its apparent encounter with the Sun.

A Basic Application of the Kepler Laws

The authentic revolution in our understanding of the movement of planets and minor bodies in the solar system came from three basic laws inferred by J. Kepler. These laws apply to bodies moving under a central force in a gravitational field. The first law indicates that the orbits of the objects in a gravitational field are elliptical with the Sun in one of the two focal points of the ellipse. The second law of Kepler states that a planet sweeps out equal areas in equal times, so the so-called areal velocity is constant. If we imagine a vector director pointing to the position of the planet, the area swept by it will be identical for the same period of time. This means that the object moving close to its perihelion will have higher angular velocity in the sky, as was nicely exemplified in the case of comets approaching to the Sun.

The third law allows astronomers to describe the physics of orbital motion and make simple, but powerful, predictions. Such law relates the orbital period *P*—in other words the time the object needs to complete one orbit—with the semimajor axis (a) of the orbit. It can be written as follows: $P^2 = k\, a^3$, where k is a constant of proportionality that is related to the mass of the body being orbited. In practice, as we can compare the orbit of a known body with another in order to infer some of its parameters, we can also use this simple equation:

$$\frac{P_x^2}{P_y^2} = \frac{a_x^3}{a_y^3} \tag{2.1}$$

In this equation the orbital period (P) and the semimajor axis (a) of two objects x, y are given. For example, if we determine the orbital period of a new asteroid from a telescopic follow-up, we could infer the semimajor axis by comparing it with that of our planet (in which we know its orbital period of one year and its mean orbital distance to the Sun can be considered its semimajor axis: 1 AU).

Using Gauss's predictions, it was von Zach himself who recovered Ceres on 7 December 1801. Among the astronomers involved in the follow-up and recovery of Ceres was Heinrich Wilhelm Olbers (1758–1840), another German astronomer working at the Bremen Observatory. The motivation for the rediscovery of Ceres prompted further observations and so on 28 March 1802 it was Olbers himself who discovered the second main-belt asteroid, now known as 2 Pallas. However, this second asteroid had a very different orbit, much more eccentric and inclined than the other planets known at the time. With the new data Herschel began to suspect that these stars could not be considered planets but a new species of smaller stars that he proposed to call asteroids. His reasoning, truly brilliant and pioneering for the time, was not accepted by the astronomical community until much later as new asteroids were discovered. In fact, the German Karl Ludwig Harding (1765–1834), following the movements of Ceres and Pallas from the Lilienthal Observatory, discovered Juno on 1 September 1804. Almost three years later, on 29 March 1807, Olbers again discovered a new asteroid, which Gauss named Vesta.

The best evidence that these objects were not planets was their small angular diameters as observed from Earth. It was not until the Dawn mission (NASA) visited and orbited Vesta that we were able to study this planetoid, a kind of surviving planetary embryo that reveals fascinating magmatic activity and a past marked by colossal impacts (Fig. 2.5).

On the basis of the discoveries made up to the first decade of the eighteenth century, Olbers developed an elementary theory that all these stars could be fragments of a larger planet that had disintegrated. I am surprised that this concept has survived through the centuries and is still an argument that can be read in books published at the end of the twentieth century. Today we know that Olbers' theory was as far-fetched as the observations of the main belt at the time were biased. Nevertheless, most of the search efforts for new asteroids in the first half of the eighteenth century were aimed at covering a strip of sky intersecting the orbital planes of these new stars. Perhaps as a result of this misdirected effort, discoveries stalled until Karl Ludwig Hencke (1793–1866), after 15 years of fruitless searching, stumbled upon asteroid 5 Astraea in 1845. In the following decades more asteroids began to appear and, as their orbits became known, many began to doubt the official ideas. It was not until 1857, when some 50 asteroids were already known, that the astronomer Francesc Aragó (1786–1853) dared to question Olbers' theory: 'The great number of objects discovered to date leads us to think that there are other causes to explain their birth.'

Knowing the ins and outs of this history, we can only praise the efforts of those early astronomers who undertook the painstaking task of discovering and tracking these objects during the eighteenth and nineteenth centuries. However, we are not really aware of the hours of work and effort invested in discovering those first hundreds of asteroids, many of them without any help other than the human eye. A real revolution would occur in the mid-nineteenth century with the birth of photography and its application to astronomy. Louis Daguerre (1787–1851), inventor of the daguerreotype, took the first images of the Moon in 1839, although they were not preserved due to a fire in his laboratory. However, the improvement of photographic plates continued after his death and astronomers realised what they could mean for the development of methods to measure the stars, without relying on costly and error-prone drawings. In fact, it was not until 1891 that the astronomer Max Wolf (1863–1932) discovered the first asteroid using the new photographic technique: 323 Brucia.

From that time we should also mention the Catalan astronomer Josep Comas i Solà (1868–1937) who, among his commendable research work on the solar system, discovered 11 asteroids, and also studied the origin of certain meteor showers or fireballs associated with falling meteorites. He began working at the Observatori Català of Rafael Patxot (1872–1964), meteorologist, businessman and tireless promoter of science and culture. Among the asteroids he first discovered was 804 Hispania in 1920. He would later discover and name 925 Alphonsina in honour of Alfonso X the Wise, 945 Barcelona, 986 Amelia, 1102 Pepita, 1117 Reginita, 1136 Mercedes, 1188 Gothlandia, 1626 Sadeya, 1655 Comas Solà and 1708 Polit. Comas Solà also made studies of comets and in fact discovered two. The first one, with a hyperbolic orbit, was called C/1925 F Schain–Comas Solà and he discovered it in 1925 near its perihelion on 6 December of that year. Subsequently, on 4 November 1926, he discovered a comet with an elliptical orbit whose orbital period is 8.83 years and which would be recognised in the catalogue of periodic comets as 32P/Comas Solá.

This was only the beginning, however, as astronomers were just beginning to glimpse the tip of the iceberg of the enormous diversity of objects, orbits and number of asteroids located in that huge storehouse we know today as the main belt. And, paradoxically enough, thousands, millions of them are still waiting to be discovered.

Main Types of Asteroids: Spectral Classes

Not all asteroids are the same. There are different populations with particular characteristics that suggest that there was a differentiated aggregation of objects as a function of distance from the Sun and also as a function of time. Even among the largest asteroids there must be important compositional differences. A clear example is the giant 1 Ceres (the number indicates the catalogue number of the *Minor Planet Center*) which, despite being about 930 km in diameter compared to the 530 km diameter of 4 Vesta, seems to be a much more primitive object that could have escaped differentiation and be a source of volatiles. Both these objects, which are shown compared to the diameter of Mars in Fig. 2.3, are the main scientific focus of NASA's Dawn mission, which aims to orbit, map and characterise them with state-of-the-art cameras and spectrometers. Curiously, Ceres is more primitive, which deepens the idea of the importance of the proportion of ices and rocks that formed these bodies. Vesta, although smaller, must have formed mostly from the aggregation of rocky planetesimals, with abundant radioactive isotopes that would efficiently heat its interior, whereas something different must have happened to Ceres. With this example we realise that we cannot lightly compare the asteroids of the solar system because they could have been born in different times and places and, to top it off, from enormously varied materials. This will be discussed in the next section.

The main belt is full of objects of a very diverse nature. Because asteroids reflect part of the light they receive from the Sun, it is possible to obtain spectra of the reflected light. By comparing such spectra, corrected and calibrated for various geometrical effects, with spectra obtained from various types of meteorites in the laboratory, it became possible to establish criteria for remote characterisation of asteroids (Johnson and Fanale, 1973; Gaffey, 1976). A review of the current status of the main associations between asteroids and meteorites was carried out by de León et al. (2011) and Trigo-Rodríguez et al. (2014).

An obvious example of the importance of comparing spectra of meteorites with those of asteroids was to identify that certain basaltic achondrites arriving on Earth in the form of three different groups called Howardites, Eucrites and Diogenites (known by the acronym HED) come from asteroid 4 Vesta. This idea has been reinforced with time and it is revealed by absorption bands characteristic of two really common minerals in these rocky materials: olivine and pyroxene (Fig. 2.4). These differentiated meteorites do not exhibit chondrules, so they are called *achondrites* and should have experienced igneous processes in a rocky planetary body with relevant magmatic activity. Certainly the parent body should still be in the main belt to be the main source of the *HED achondrites*. Pieters and McFadden (1994) provided an excellent review of this affinity between HED and 4 Vesta achondrites that has been accepted to the present day. Asteroids exhibiting similar reflectance spectra are known as Vestoids, and they are explained due to being released from 4 Vesta by colossal impacts that still mark its surface (Fig. 2.5).

We could think that distinctive features allow remote identification through asteroid and meteorite spectra, but unfortunately this is not the case. For example, we also know that certain asteroids of spectral class S have clear similarities to ordinary chondrites, but there are countless asteroids sharing these reflective properties. This is a consequence of the fact that there have been several fragmentations over the eons of the parent asteroids of ordinary chondrites. Nesvorný et al. (2002a) recently discovered that there is a family of 38 asteroids dynamically associated with asteroid 832 Karin that must have been produced by the fragmentation of a larger asteroid about 5.8 Ma ago.

Many processes have altered the initial surfaces of the asteroids over the eons, producing impact gardening, and at the same time the surfaces are exposed to irradiation by solar light and cosmic rays. All these processes produce significant changes in the reflectance properties of these asteroids in a process known as *space weathering*. In addition, to explain the complexity of the asteroid regions, there exist different populations and families. All this complexity is reminiscent of archaic processes, often stochastic, that gave rise to these objects and families (Michel et al., 2003). Dynamic evolution has caused them to move from their regions of origin and, in addition, impact processing was intense for most of them, both destroying asteroids and producing heterogeneous fragments that complicate the scenario (Walsh et al., 2012). These and other factors make the main asteroid belt diverse and dynamically complex due to the introduction of chaos associated with the so-called orbital resonances, which we will discuss in another section.

Returning to the fact that there are asteroids with similar mineralogical properties forming populations in the main belt, I would like to give some examples. It is well known that enstatite asteroids are found in an inner region close to Mars. Not surprisingly, their particular mineralogy and oxygen isotope ratios, as measured in the rare enstatite chondrites, lead us to suspect that they are the major progenitor building blocks of the Earth. They are possibly genuine fossils of a virtually extinct population that gave rise to the terrestrial planets by aggregation in the past. Only those asteroids of this nature that were dynamically stored in this region were able to survive. Also, asteroids of spectral class S, which equates to common ordinary chondrites, dominate the interior of the solar system. Then, asteroids of carbonaceous nature and spectral class

C are more frequent around the centre of the main belt. Finally, the presumably more primitive D and P spectral classes associated with the more volatile-rich carbonaceous chondrites dominate the outer region. We would expect this if these objects were an amalgam of high-temperature solid materials (such as silicates or metals formed in the inner protoplanetary disc) and low-temperature ones (such as ices or organic matter) from the cooler outer regions. This is consistent, on the other hand, with the notion that the protoplanetary disc was subject to a temperature gradient throughout its early evolution (Wasson, 1985; Trigo Rodríguez, 2012a).

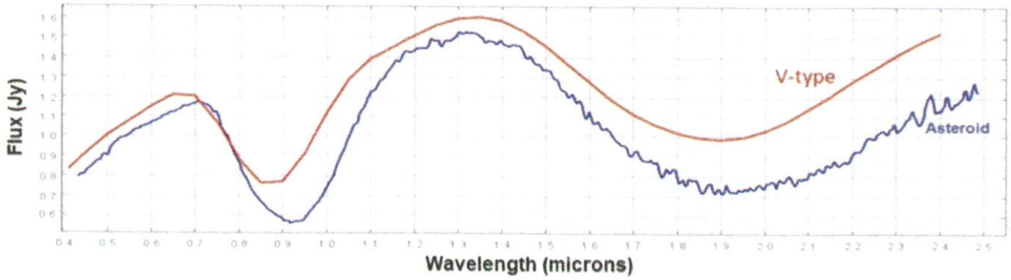

Fig. 2.4 The reflectance spectrum of a Vestoid (in blue) compared with the average shape of the V-type

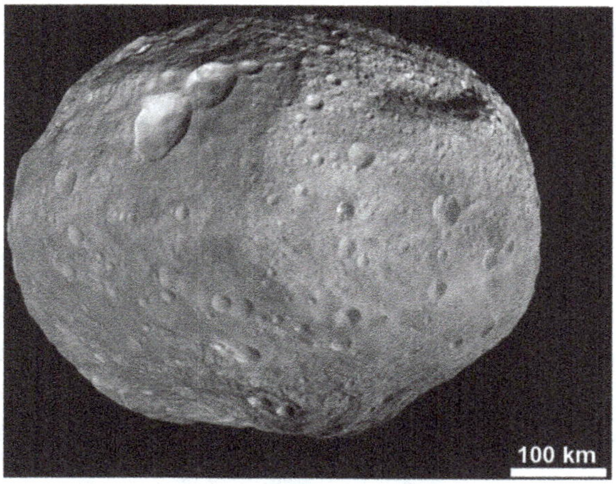

Fig. 2.5 Asteroid 4 Vesta mapped by NASA's Dawn probe (NASA/JPL-Caltech/UCLA/MPS/DLR/IDA)

2.3 Resonances as Sources of Meteorites

In recent decades there has been enormous progress in our understanding of the dynamic mechanisms that send both main-belt asteroids into the NEO near-Earth region and the smaller fragments or meteoroids that most frequently encounter our planet and produce meteorites.

Large asteroids are subject to collisions, which until relatively recently were thought to be the cause of sending asteroid fragments into resonances. However, we now know that it is the fragments produced by a collision and that are smaller than about 20 km in diameter that are mostly subject to the so-called Yarkovsky effect. This is a non-gravitational effect in which the surface of asteroids absorbs sunlight and radiates it back with a certain delay. This radiated energy is negligible, but because it is a constant process over time, it creates a force opposite to the motion that contributes to the asteroid losing potential energy and falling towards the Sun, one revolution after the other. As they decay in their orbit, these small asteroids will cross so-called mean motion resonances with the various planets. Among the most intense are those associated with Jupiter, Saturn and Mars. For the sake of simplicity, I will only discuss some of them.

The most intense mean motion resonances with Jupiter are: 3:1, 4:1 and 5:2, as shown in Fig. 2.6. I will give an example to explain the origin of this double numbering to name them. Thus, the 3:1 resonance (read: three, one) occurs when the asteroid completes three orbits while Jupiter makes one. As the asteroid falls into these regions, its orbit, affected by Jupiter's gravitational influence, can acquire such an eccentricity that it is thrown into the near-Earth region. In this region it may orbit for a few million years and then return to the main belt (Morbidelli et al., 2002). A simple orbital diagram shows that the distribution of asteroids as a function of their distance from the Sun has forbidden regions called Kirkwood gaps (Fig. 2.6). Note that in Fig. 2.6 such 'gaps' are in the form of forbidden lines for the presence of objects in the source region of the *main belt.* Some of these regions that launch asteroids towards the terrestrial planets, such as the 3:1 mean-motion resonance of

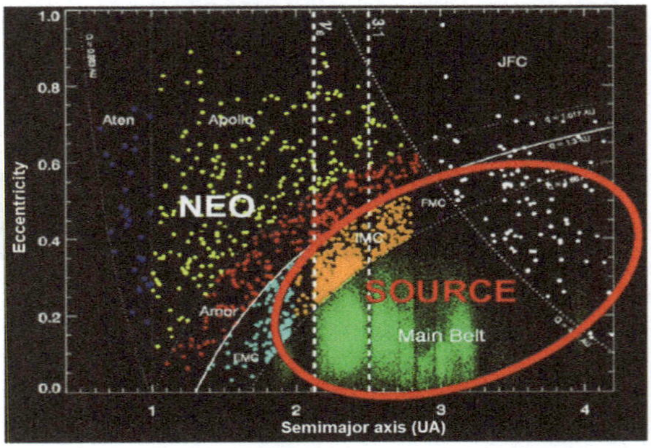

Fig. 2.6 Diagram of the semimajor axis versus eccentricity of objects in the inner solar system. Each point represents an object at the present time (image courtesy of P. Michel/Marco Polo-R team/ESA)

asteroids with the giant planet Jupiter or the secular resonance (ν6) with the planet Saturn, are also marked in dashed lines in this figure. Figure 2.6 also indicates the main populations of NEOs (Apollo, Aten and Amor), as well as other comet regions such as Jupiter Family Comets (JFCs) and Near-Earth Comets (NECs).

As we have seen, other important resonances are the so-called secular resonances. These resonances occur when the precession of two orbits is synchronised. As a consequence, the asteroid falling into the resonance ends up completing precession at the same time as the planet and this causes extreme changes in the asteroid's eccentricity and inclination. The most important secular resonance is the so-called ν6 resonance that asteroids undergo due to the influence of the planet Saturn (see Fig. 2.6). Asteroids that dare to approach this region undergo a change in their eccentricity that causes them to dangerously cross the orbit of Mars. It is precisely in these close encounters that such asteroids can be ejected from the main belt into the innermost regions. The study of asteroid orbits in the near-Earth region and simulations indicate that this resonance is one of the most important sources of NEOs (Nesvorný et al., 2002b).

The transport mechanisms of rocks that are detached in the collisions or disintegration of asteroids and comets to Earth are not nearly as fast as we might think (Morbidelli et al., 2002). Rocks with a diameter between one metre and one thousandth of a millimetre are called *meteoroids*, while from one metre upwards they are small *asteroids*. On average, such rocks usually spend several tens of millions of years in interplanetary space. These times are known thanks to the cosmic-ray exposure ages (CREAs), which I will explain in more detail in Chap. 5. Basically, depending on the time of exposure to interplanetary space since they have been detached from their parent asteroid, these rocks accumulate certain isotopes, and subsequent study of their abundances once they arrive on Earth allows us to date the time since they were detached from their parent body.

Our increased understanding of dynamical resonances means that we know their transport mechanisms. A series of cosmic ricochets will allow these rocks to meet our planet on their way and, only in the most favourable geometrical conditions, with a suitable velocity and angle of incidence, will they allow the survival of debris that will reach the Earth's surface. Such objects are called meteorites. Typically, more than 95% of the mass of these rocks in space will be lost to the atmosphere in the luminous *fireball* phase, producing fine dust, hot gas and a small percentage of light, enough to make these phenomena extraordinarily striking to the lucky casual observer. Most of the meteorites that reach the Earth's surface are rocks composed essentially of silicates, although there are also metallic ones, composed mainly of iron and nickel alloys. We know that most of these meteorites come from rocks, individually propelled from the main belt. However, as I will explain in later chapters, we have been collecting evidence that certain near-Earth objects can produce meteorites.

2.4 Transitional Objects and Comets

An interesting type of object is composed of asteroids that have nevertheless shown or are showing some kind of cometary activity. They are found in the outermost part of the main asteroid belt where they have been recently identified. In the outer region of the belt, certain objects formed by the aggregation of rocky components, organic matter and ices would have been stored. Such objects would have retained certain amounts of ice, which would sublimate under certain conditions producing the observed activity, but, like the asteroids, would be mostly composed of silicates (Hsieh and Jewitt, 2006). The best known of these is the transitional comet-asteroid object 133P/Elst-Pizarro, which in fact went unnoticed as an asteroid for decades, so it also has the catalogue number 7968 (Fig. 2.7).

It should be made clear to the reader that the word *comet*, which is identical in many modern languages, comes from the Greek word for *hair*. In fact, if comets differ in any way from other stars in the sky, it is because they are accompanied by long hairs, known as *tails*. These tails are produced when comets approach the Sun, when the ices they contain sublimate and the gases they release propel small particles of ice and dust. This behaviour is possible because comets are extremely fragile bodies, having formed from the aggregation of small solid particles and ices in the outermost region of the protoplanetary disc. In fact, they are extraordinarily porous bodies whose average density is less than that of water. It seems that low-velocity collisions of the tiny materials formed a spongy, highly porous structure that still characterises them today.

In addition, there are some transitional objects that behave like comets stored in the main belt. This is because they can develop a faint coma or tail, even if only

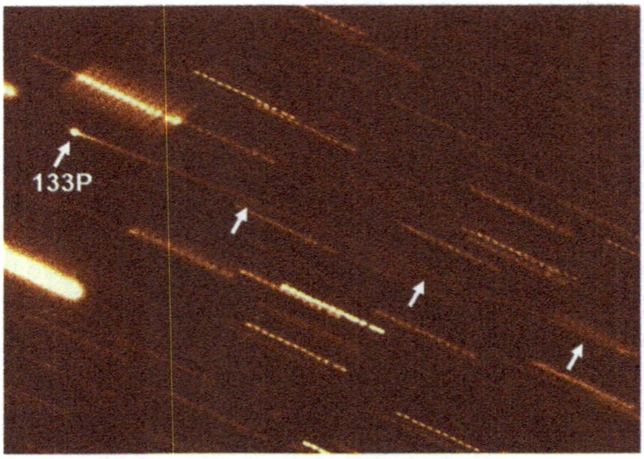

Fig. 2.7 The main-belt comet 133P/Elst-Pizarro recorded on 7 September 2002 with the Mauna Kea telescope at the University of Hawaii. In this image the comet has been tracked so that the stars appear as arcs. The comet is indicated by its numbering and successive arrows follow its dusty tail in the star field (courtesy of D. Jewitt and H. Hsieh/U. Hawaii)

sporadically in some cases. They appear to be part of a population of thousands of asteroids, many of them in a *collisional family* called the asteroid 24 Themis family. On the other hand, the discovery that the asteroid (101,955) Bennu visited by NASA's OSIRIS-REx probe has a cloud of fragments aroused curiosity, although it has been proposed that they must have been ejected by small impacts on its surface (Bottke et al., 2020).

Although there are active asteroids, most asteroids do not show cometary activity, suggesting that the proportion of ices on these bodies is significantly lower than on Trans-Neptunian objects (TNOs) and comets, or at least that they can be buried in regolith. Obviously, comets must be covered with abundant regolith due to the landscaping of the surface as a consequence of countless impacts over the eons. Among the active asteroids, it has been modelled that a chance impact was the most reasonable mechanism for the emission of materials from the surface of the asteroid Sheila.

The existence of main-belt comets indicates that the transition between rocky and icy objects, the former formed close to the Sun and the latter far from it, is not sudden, but that there is a region about three times the Sun–Earth distance populated with objects of intermediate composition. The region in which they formed would have incorporated organic matter, hydrated minerals and ices. On the other hand, it is suspected to be a population of high cosmochemical interest given that the Earth has been impacted by many asteroids. If these do indeed have some water content, they could have contributed to the formation of the terrestrial oceans around 3,900–3,800 Ma, the time when the migration of the giant planets has been postulated (Gomes et al., 2005). Such an inward migration would have gravitationally dispersed thousands of transitional objects into collision paths with the terrestrial planets and would naturally explain the Late Heavy Bombardment that has been inferred from the age of large lunar basins and craters based on the study of the age of lunar rocks collected by the Apollo missions, and the lunar meteorites (Trigo-Rodriguez, 2012b). The chemical patterns found in Apollo samples permitted the meteorite researchers to distinguish lunar meteorites, the first being Allan Hills A81005, an anorthositic breccia found in Antarctica in 1982.

2.5 Comets: The Trans-Neptunian Belt and the Oort Cloud

The fact that comets have preserved their primordial properties is due to the fact that most of them have remained stored in distant regions of the solar system without being heated by radioactive decay or sunlight, as happened to asteroids. The fascination that has always surrounded these icy objects has been reflected in the literature of even primitive civilisations. Based on the study of the different appearances of comets, it was the Briton Edmond Halley (1656–1742) who first identified the periodicity of these objects in their visibility from Earth. The first comet he identified, whose spectacular orbital period of about 76 years made it a remarkable object, was named comet 1P/Halley. The fascination with learning more details about this comet led different space agencies to design a whole armada of spacecrafts to closely explore the nucleus of comet 1P/Halley, usually hidden in the sublimated gas coma of its interior. The plethora of results obtained by these probes that passed through the comet's coma in 1986 was a first breakthrough in the enigmatic origin of these objects, although many other questions were raised. In recent decades, the scientific community has been accumulating evidence that these are primitive objects that preserve within them the keys to the physico-chemical processes that took place during the origin of the solar system. To corroborate these ideas, NASA's *Stardust* mission in 2006 recovered for the first time particles from a comet, catalogued as 81P/Wild 2, which are in fact the first solid samples recovered in situ from a solar system object other than the Moon (Brownlee et al., 2006). The studies of that probe revealed that Comet 81P/Wild 2 has a diameter of 4.5 km and is mostly composed of a mixture of ices, organic matter and mineral grains whose typical size is about 10 μm (a micrometre is one thousandth of a millimetre) (Fig. 2.8).

The Oort Cloud and the Trans-Neptunian belt regions are among the most populated in the solar system, containing of the order of 10^{12} objects. These distant bodies are basically composed of ice, organics and fine-grained silicates. These poorly processed silicates that have remained in a region far from the Sun are very cold, 'frozen', so the materials that form them are the most primitive we can find. In fact, these objects are the precursors of comets, bodies that escape from the Cloud and the Belt and enter the region close to the Sun due to gravitational perturbations. From the *Oort Cloud*, comets arrive in long-period orbits reaching tens of thousands of years (Fig. 2.9). Objects in the Cloud can only be detected, with current technology, if they transform into comets. No bodies are observable there given their relatively small size and the enormous distances at which they are located. Trans-Neptunian

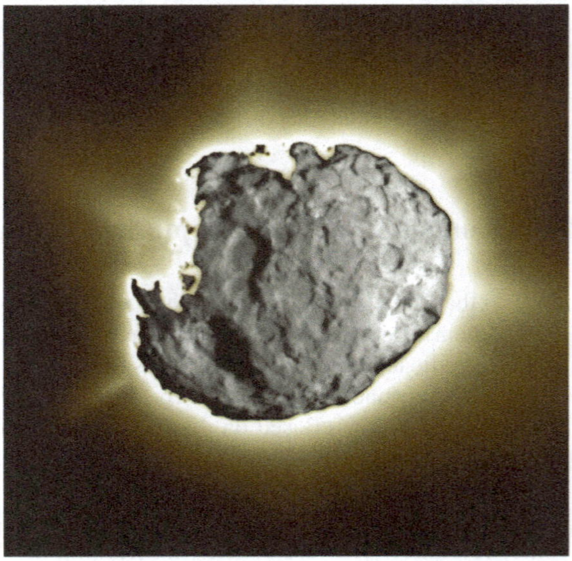

Fig. 2.8 Comet 81P/Wild 2 surface image overexposed to show the active regions that emit the sublimated gas and drive the tiny particles brought to Earth by the Stardust probe (Stardust/NASA)

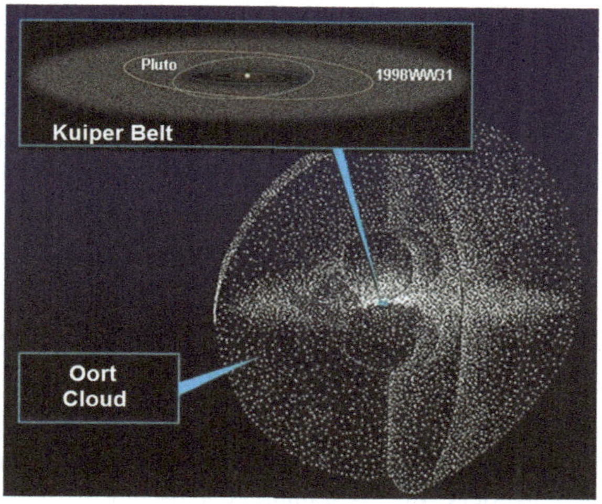

Fig. 2.9 Diagram to show the Kuiper Belt and the Oort Cloud. Just as a reference, the upper inset shows the orbits of Pluto and 1998 WW31, a non-resonant TNO located in the outermost region of the Kuiper Belt (adapted with permission from D. K. Yeomans) (JPL/NASA)

objects (TNOs) are the 'parents' of the so-called Centaurs but also of short-period (<20 years) comets (Jewitt, 2008). Fortunately, much progress has been made in the knowledge of these objects in the last decade thanks to dedicated surveys. Among the most notable events related to this population, the discovery in 2005 of three TNOs of similar size to Pluto should be highlighted: Eris, Makemake and Haumea (see Table 2.2).

The discovery of these large TNOs marked a new era in our understanding of the outer solar system. Eris and Makemake were discovered on 21 October 2003 and 31 March 2005, respectively, by a team led by Michael E. Brown, Chadwick Trujillo and David Rabinowitz of the California Institute of Technology (Caltech) at Palomar Observatory. Haumea was discovered at the Sierra Nevada Observatory by an Instituto de Astrofísica de Andalucía (CSIC) group led by José Luis Ortiz, who has made important contributions to a better understanding of the dimensions and physical properties of TNOs using stellar occultations (Sicardy et al., 2011; Ortiz et al., 2012). Based on the results of these studies in which the professionals

Table 2.2 Major Trans-Neptunian objects with a diameter greater than 1,000 km

Object	Diameter (km)
(136,199) Eris	2,326
(134,340) Pluto	2,340
(90,377) Sedna	1,600
(136,472) Makemake	~1,500
(136,108) Haumea	1,270
Charon (Pluto's satellite)	1,210
(84,522) 2002 TC302	1,150
(50,000) Quaoar	1,140

recognise the valuable task of amateur observatories, Table 2.2 gives the diameters of the largest TNOs known so far. Different surveys are now concentrated in obtaining new discoveries in these remote regions of our planetary system, and the search for the so-called planet X is helping in that regard. Caltech astronomers Konstantin Batygin and Mike Brown announced in 2015 that they have found mathematical evidence suggesting the presence of a Neptune-sized planet in a highly elongated orbit about 20 times farther from the Sun than Neptune. For now it is just a theoretical object, but who knows if it is just in another populated region awaiting discovery?

Questions for Chapter 2

Q.1. Using the third law of Kepler, and knowing that an asteroid orbits the Sun with a semimajor axis of 2 AU, try to estimate its orbital period. [Solution: 2.82 terrestrial years.]

Q.2. The current asteroids are real survivors of large impacts. The initially porous planetesimals were subjected to impacts that compacted and heated to some extent their interiors. Models of impact compaction of 100 km-sized asteroids indicate that the continuous bombardment of km-sized impactors have sculpted these asteroids, transforming most of them into shattered bodies called rubble-piles. Beitz et al. (2016) found that from the size of the largest impact crater on the asteroid surface, the thickness of the regolith layer can be inferred. By using such an approach try to predict the possible thickness of the regolith of asteroid Mathilde, which is 53 km in diameter. [Solution: The largest craters such as Ishikari and Karoo are about 30 km in diameter, so probably the entire body is heavily fractured.]

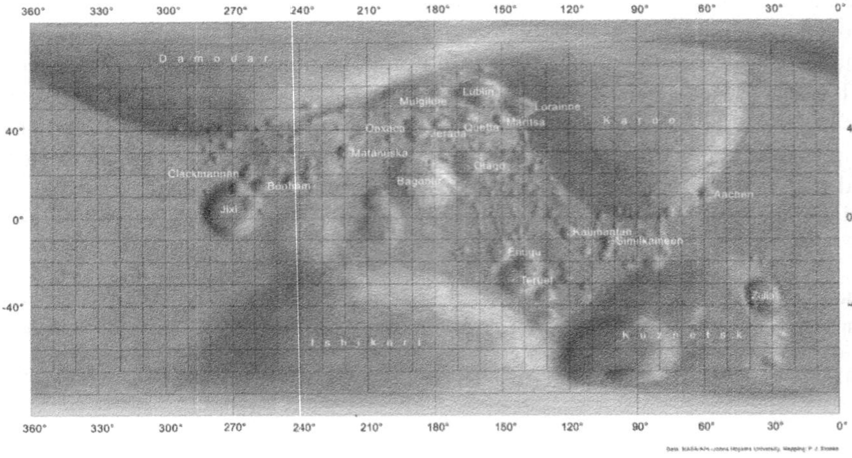

Figure for Q.2. Cylindrical map of the craters on asteroid Mathilde (NASA/APL-Johns Hopkins University, Mapping: P.J. Stooke).

Q.3. Dynamical studies have revealed that the entire NEO population is replenished in a few million years. Current NEOs will leave the near-Earth region, and other main belt objects will occupy their places. This is due to the orbital evolution of the main asteroid belt bodies, largely because of the dynamic effects induced by gravitational resonances with Jupiter, and Saturn. (a) How many times has the near-Earth region been approximately replenished since the Late Heavy Bombardment occurred about 3.8 Gyr ago? and (b) What can you conclude about the hazard associated with NEOs? [Solution: (a) At least the entire population has been completely renovated 1,000 times. (b) The survival of our civilization is linked with the need to monitor closely the NEOs, a highly changing population of bodies.]

Q.4. New long-period comets are being discovered all the time. A recent study reveals that during the pre-telescopic era our ancestors were able to discover about one new comet every year (Hughes, 2003). Obviously now, using ground-based telescopic surveys or space telescopes and sensors, we discover tens of such objects every year. According to these numbers you can roughly quantify the magnitude of the number of icy objects stored in the Oort Cloud. Some of these objects come back to these remote regions still under the gravitational influence of the Sun, but others shorten their orbits and finally decay into the inner solar system. [Solution: Considering a similar rate of comet appearances over the last 3,500 Myr (end of the Late Heavy Bombardment), a minimum of 35,000 million comets has visited us since then. Being probably a lower limit, it exemplifies well the magnitude of icy bodies contained in the Oort Cloud.]

Online Resources

Comet Catalog: http://www.aerith.net/comet/catalog/index-periodic.html
Minor Planet Center: https://minorplanetcenter.net//

References

Beitz E., J. Blum, M.G. Parisi and J.M. Trigo-Rodríguez (2016) The collisional evolution of undifferentiated asteroids and the formation of chondritic meteoroids. *Astrophysical Journal*, in press, 29 pp.

Bottke, W.F. et al. (2020) Meteoroid impacts as a source of Bennu's Particle Ejection Events. *Journal of Geophysical Research: Planets* 125, Issue 8, article e06282

Brown P., R.E. Spalding, D.O. Revelle, E. Tagliaferri and S.P. Worden (2002) The flux of small near-Earth objects colliding with the Earth. *Nature*, 420, 294–296.

Brown P. et al. (2013) A 500-kiloton airburst over Chelyabinsk and an enhanced hazard from small impactors. *Nature*, 503, 238–241.

Brownlee et al., 2006.Brownlee D. et al. (2006) Comet 81P/Wild 2 under a microscope. *Science* 314, 1711–1716.

De León, J., J. Licandro, M. Serra-Ricart, N. Pinilla-Alons and H. Campins (2011) Finding the compositional diversity of the Solar System. In *Highlights of Spanish Astrophysics VI, Proceedings of the IX Scientific Meeting of the Spanish Astronomical Society (SEA)*, M.R. Zapatero Osorio et al. (eds.), pp. 555–569.

Eugster O. et al. (2006) Irradiation records, cosmic-ray exposure ages, and transfer times of meteorites. In *Meteorites and the Early Solar System II*, D. Lauretta, H.Y. McSween (eds.), The University of Arizona Press, Tucson, AZ, USA, pp. 829–851.

Gaffey M.J. (1976) Spectral reflectance characteristics of the meteorite classes. *Journal of Geophysical Research* 81, 905–920.

Gomes R., H.F. Levison, K. Tsiganis and A. Morbidelli (2005) Origin of the cataclysmic Late Heavy Bombardment period of the terrestrial planets. *Nature* 435, 466–469.

Grady, 2000.Grady M. (2000) *Catalogue of Meteorites*. Cambridge Univ. Press, Cambridge, UK.

Gutiérrez Buenestado P.J. (2012) Asteroides y Cometas, Colección ¿Qué sabemos de... Editorial Catarata-CSIC, Madrid.

Hsieh and Jewitt, 2006.Hsieh H. and D. Jewitt (2006) A population of comets in the Main Asteroid Belt. *Science* 312, 561–563.

Hughes D.W. (2003) Early long-period comets: their discovery and flux. *Monthly Notices of the Royal Astronomical Society* 339, 1103–1110.

Jewitt D. (2008) Kuiper Belt and Comets: An observational perspective. In *Trans-Neptunian Objects and Comets*, Swiss Society for Astrophysics and Astronomy, Springer-Verlag, Heidelberg, Germany, pp. 1–78.

Johnson T.V. and F.P Fanale (1973) Optical properties of carbonaceous chondrites and their relationship to asteroids. *Journal of Geophysical Research* 78, 8507–8518.

Michel P., W. Benz, D.C. Richardson (2003) Disruption of fragmented parent bodies as the origin of asteroid families. *Nature* 421, 608–611.

Morbidelli et al., 2002.Morbidelli A., W. F. Bottke Jr., C. Froeschlé, and P. Michel (2002) Origin and evolution of Near-Earth Objects. In *Asteroids III*, W.F. Bottke, A. Cellino, P. Paolicchi and R.P. Binzel (eds.), The University of Arizona Press, Tucson, AZ, USA, pp. 409–422.

Nesvorný D., W.F. Bottke Jr., L. Dones and H.F. Levison (2002a) The recent breakup of an asteroid in the main-belt region. *Nature* 417, 720–722.

Nesvorný D., S. Ferraz-Mello, M. Holman and A. Morbidelli (2002b) Regular and chaotic dynamics in the Mean-motion resonances: Implications for the structure and evolution of the asteroid belt. In *Asteroids III*, W.F. Bottke, A. Cellino, P. Paolicchi and R.P. Binzel (eds.), The University of Arizona Press, Tucson, AZ, USA, pp. 379–394.

Nittler L.R., R.M. Stoud, J.M. Trigo-Rodríguez, B.T. De Gregorio, C.M.O.'D. Alexander, J. Davidson, C.E. Moyano-Cambero and S. Tanbakouei (2019) A cometary building block in a primitive asteroidal meteorite. *Nature Astronomy*, doi: https://doi.org/10.1038/s41550-019-0737-8, 8 pp.

Ortiz J.L., F.J. Aceituno, J.A. Quesada, J. Aceituno, M. Fernández, P. Santos-Sanz, J.M. Trigo-Rodríguez, J. Llorca, F.J. Martín-Torres, P. Montañés-Rodríguez and E. Pallé (2006) Detection of sporadic impact flashes on the Moon: Implications for the luminous efficiency of hypervelocity impacts and derived terrestrial impact rates. *Icarus* 184, 319–326.

Ortiz J.L. et al. (2012) Albedo and atmospheric constraints of dwarf planet Makemake from a stellar occultation, *Nature* 491, 566–569.

Pieters C.M. and L.A. McFadden (1994) Meteorite and asteroid reflectance spectroscopy: Clues to early solar system processes. *Annual Review of Earth and Planetary Sciences* 22, 457–497.

Rubin, 2002.Rubin A.E. (2002) *Disturbing the Solar System*. Princeton University Press, Princeton, NJ, USA, 361 pp.

Sicardy B. et al. (2011) A Pluto-like radius and a high albedo for the dwarf planet Eris from an occultation. *Nature* 478, 493–496.

Tanbakouei S., J.M. Trigo-Rodríguez, J. Blum, I. Williams and J. Llorca (2020) Comparing the reflectivity of ungrouped carbonaceous chondrites with that of short period comets like 2P/Encke. *Astronomy & Astrophysics* 641, A58, 7 pp.

Trigo-Rodríguez J.M. (2012a) *Meteoritos, Colección ¿Qué sabemos de...* Editorial Catarata-CSIC, Madrid.

Trigo-Rodríguez, 2012b.Trigo-Rodríguez J.M. (2012b) Las raíces cósmicas de la vida. Colección El Espejo y la Lámpara, Servei Publicacions UAB, Barcelona.

Trigo-Rodríguez J.M. (2015) Aqueous alteration in chondritic asteroids and comets from the study of carbonaceous chondrites. In *Planetary Mineralogy, European Mineralogical Union Notes in Mineralogy* 15, pp. 67–87.

Trigo-Rodríguez J.M. and I.P. Williams (2017) Dynamic sources of contemporary hazard from meteoroids and small asteroids. In 'Assessment and mitigation of asteroid impact hazards', Springer, *Astrophysics and Space Science* 46, pp. 11–32.

Trigo-Rodríguez J.M. and J.M. Madiedo (2013) Asteroids and comets as progenitor bodies of meteorites: Their danger of impact with the Earth. *Revista Enseñanza de las Ciencias de la Tierra-AEPECT* 21–3 (ISSN: 1132–9157), Edita Asociación Española para la Enseñanza de las Ciencias de la Tierra, Madrid, pp. 293–301.

Trigo-Rodríguez J.M., E. Lyytinen, D.C. Jones, J.M. Madiedo, A.J. Castro-Tirado, I. Williams, J. Llorca, S. Vítek, M. Jelínek, B. Troughton and F. Gálvez (2007) Asteroid 2002NY40 as a source of meteorite-dropping bolides. *Monthly Notices of the Royal Astronomical Society* 382, 1933–1939.

Trigo-Rodríguez J.M., C.E. Moyano-Cambero, J. Llorca, S. Fornasier, M.A. Barucci, I. Belskaya, Z. Martins, A.S. Rivkin, E. Dotto, J.M. Madiedo and J. Alonso-Azcárate (2014) UV to far-IR reflectance spectra of carbonaceous chondrites – I. Implications for remote characterization of dark primitive asteroids targeted by sample-return missions. *Monthly Notices of the Royal Astronomical Society* 437, 227–240.

Trigo-Rodríguez, J.M., A. Rimola, S. Tanbakouei, V. Cabedo-Soto and M.R. Lee (2019) Accretion of water in carbonaceous chondrites: Current evidence and implications for the delivery of water to early Earth. *Space Science Reviews* 215(18), 27 pp.

Walsh K.J., D.C. Richardson, P. Michel (2012) Spin-up of rubble-pile asteroids: Disruption, satellite formation, and equilibrium shapes. *Icarus* 220, 514–529.

Wasson J.T. (1985) *Meteorites: Their Record of Early Solar System History*. W.H. Freeman and Company, New York, USA.

Chapter 3
The Origin and Nature of Comets

Since ancient times, the sudden and unexpected appearance of comets in the firmament has aroused all kinds of fears and superstitions in the human race. Even some of the most favourable interpretations showed how the changing appearance of the cometary coma and the tail was reminiscent of a kind of cosmic sword trying to cut the Earth to pieces. With the passage of time, and with more enlightened views, it was nonetheless not easy to interpret the structure and varying appearance of these bodies, so for centuries they were thought to be atmospheric phenomena. Several millennia of written human history had to pass before their nature was understood.

The fascination that has always surrounded the observation and tracking of comets has been reflected in the literature of even primitive civilisations. The word comet comes from the Greek 'coma', which means hair. The changing shape and rapid movement of these bodies night after night attracted the attention of astrologers who wished to predict the behaviour of the heavens. Comets resembling a dagger or sword often served as an excuse to predict disasters and even precipitated the fall of the throne of kings. As unparalleled astronomical objects, their variable appearance and luminosity made them too attractive and frightening for most of our ancestors. Their interpretation transcended the contemplation of the firmament and became rooted in myths and legends about the influence of the cosmos on mankind. Even today, when we know their nature quite well, bright comets arouse enormous scientific interest and social curiosity.

Based on the study of the various comet apparitions, it was Edmond Halley (1656–1742) who first became aware of the periodicity of the visits of these objects, often elusive in their visibility as seen from Earth. He used historical records of comet appearances that he rightly interpreted as a valuable legacy over generations. On the basis of these records, Halley understood that one famous comet, whose spectacular nature made it a remarkable object virtually every time it passed close to the Sun, must follow an elliptical orbit with an orbital period of about 76 years. In honour of this pioneer, this comet was called Halley's Comet. Today it is the most famous and well-known comet, which in 1986 led to the efforts of a flotilla of space missions to study its environment and nature. Among them, the European Space Agency (ESA) sent the

J. M. Trigo-Rodríguez, *Asteroid Impact Risk*, Impact Studies,
https://doi.org/10.1007/978-3-030-95124-5_3

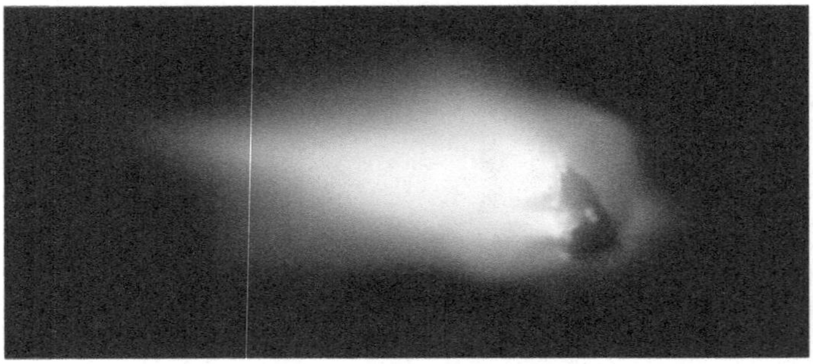

Fig. 3.1 Comet 1P/Halley recorded by the Halley Multi-Colour Camera (HMC) on board the Giotto probe (ESA/MPAe Lindau)

Giotto probe, named in memory of the famous Florentine painter Giotto di Bondone (1267–1337) who, after seeing Halley's Comet on its apparition in 1301, captured it in his painting of the birth of Christ. The Giotto probe recorded the first detailed images of the comet's nucleus (Fig. 3.1). Given the spectacular nature of some of its apparitions, it is not surprising that this comet became so famous. The fact is, however, that there are many others of shorter period and much more frequent visitation. All of them are catalogued by the *Minor Planet Center (MPC),* an international organisation dedicated to cataloguing asteroids and comets, using the order number and establishing a name or, at least, a preliminary alphanumeric designation. In the case of comets, the surname of the discoverer(s) is usually attributed, so of course the first object in the MPC catalogue of periodic comets is 1P/Halley (Table 3.1).

Comets like 1P/Halley are usually relatively fragile bodies, stacked in blocks hundreds of metres across and generally only a few kilometres in diameter. They are made up of smaller blocks loosely bound together by an amalgam of ice, organic matter and small rocky minerals. Because of their highly volatile component, many short-period comets experience a continuous loss of mass on successive approaches to the Sun. This gradually wears away and weakens the materials that act as a link between the forming blocks and so fracture occurs, often accompanied by an abrupt increase in the luminosity of the ensemble as a consequence of the exposure to solar radiation of these friable components (Trigo-Rodríguez et al., 2010). Continuous observations of these objects since the nineteenth century have revealed that they frequently disintegrate into several pieces. Such fragmentations usually produce multiple fragments, some of which can be solid blocks of tens or hundreds of metres that do not show cometary activity. This is the origin of so-called 'complexes', families of objects that must have arisen from the fragmentation of a parent body.

Other comets are very resilient and periodically surprise astronomers with huge outbursts in which they increase their magnitude of brightness. Fascinated by these events, I have been able to conduct several studies to investigate the properties and nature of such processes in which they suddenly increase their luminosity, pointing to their origin in processes of sublimation of their ices (Trigo-Rodríguez et al., 2007, 2008a, 2010). In such events comets emit significant amounts of dust that, under

Table 3.1 Famous periodic comets with their main dimensions, estimated density, orbital period, class (Halley type: HTC; Encke type: ETC; Jupiter family: JFC; or main-belt asteroid: MBA). Dimensions are accurate if there has been a spacecraft visit or otherwise are remotely estimated. Density estimates can have a minimum of 25% error. (Various literature sources, including: Boehnhardt et al. (1999); Blum et al. (2006); Jewitt (2004), Schambeau et al. (2015).)

Comet	Dimensions/estimated diameter (km)	Average density (g/cm^3)	Orbital period (years)	Eccentricity	Class
1P/Halley	15 × 8	0.,6	75.32	0.967	HTC
2P/Encke	2.4	0.8	3.3	0.848	ETC
7P/Pons–Winnecke	520	3.5	6.37	0.634	JFC
17P/Holmes	3.4	0.9	6.89	0.432	JFC
19P/Borrelly	8 × 4 × 4	0.3	6.86	0.623	JFC
26P/Grigg–Skjellerup	1.5	1.0	5.31	0.633	JFC
29P/Schwassmann–Wachmann 1	60 ± 7	–	14.68	0.045	JFC
67P/Churyumov–Gerasimenko	4.1 × 3.3 × 1.8 2.6 × 2.3 × 1.8	0.5	6.45	0.641	JFC
73P/Schwassmann–Wachmann 3	1.1	–	5.44	0.686	JFC
133P/Elst–Pizarro/(7968)	3.8	1.3	5.63	0.157	MBA

the very small gravitational field, is expelled by the pressure of the sublimated gas and expands and fragments in the interplanetary medium. This process generates the so-called coma. These are faint envelopes of micrometre-sized particles that scatter sunlight, creating a blurry appearance due to the gradually expanding envelope. Comet 17P/Holmes in particular experienced an abrupt and powerful outburst in which it went from being barely visible at stellar magnitude +17–+2.5, representing almost a million-fold increase in brightness. This amazing discovery was made by Canarian amateur A. Henriquez Santana. Just one hour after his observation, the object was high enough for our all-sky camera located at the Observatori Astronòmic del Montsec (OAdM) to capture it at magnitude +8 and increasing its brightness to magnitude +6 along the night of 2007 October 23/24. Our follow-up of the evolution of the outburst with several instruments, including the Carlos Sánchez infrared telescope of the Instituto de Astrofísica de Canarias (IAC), allowed us to measure the expansion of the envelope as shown in Fig. 3.2 (Trigo-Rodríguez et al., 2007, 2008a).

Another comet that experiences periodic outbursts is 29P/Schwassmann–Wachmann 1. Our observations over almost 20 years suggest that the ices of this comet sublimate and produce outbursts that increase its brightness by up to five magnitudes. Our photometric monitoring using different filters clearly indicates that the magnitude of the comet's coma increases as the centimetre-sized particles fragment in their expansion and transform into micrometre-sized particles that are able to scatter sunlight (Fig. 3.3).

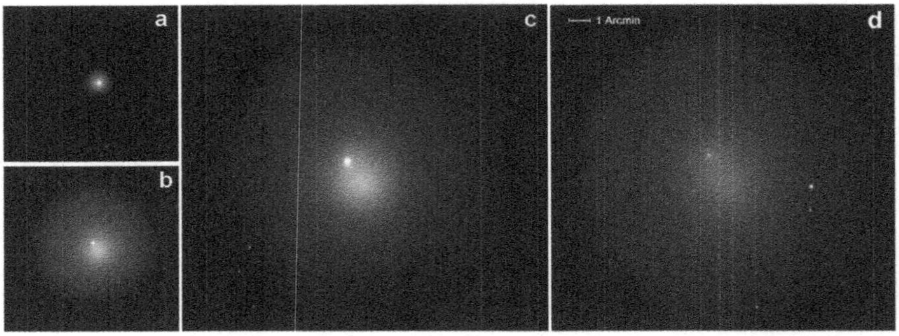

Fig. 3.2 Images of Comet 17P/Holmes in expansion after the outburst obtained with R filter by Albert Sànchez from the Observatori de Gualba (MPC 442) using a SC 36.0 telescope at f/6. Image **c** is taken by the author with the infrared telescope Carlos Sánchez (IAC). The sequential expansion of the envelope after the outburst experienced by the comet can be seen in images taken in 2007: **a** Oct. 25.01; **b** Oct. 27.86; **c** Oct. 30.1; and **d** Nov. 1.01. Dates include decimal days (image adapted from Trigo-Rodríguez et al.,2008b)

Only a few decades ago we thought of comets as fragile snowballs and assemblages of organic matter formed in the outer regions of the solar system. Some of them might fit these criteria, but others might be objects formed in inner regions even though they were gravitationally scattered by the planets towards the outer reaches of our planetary system. There is no doubt that our view of these bodies has changed radically thanks to the space missions that both the European Space Agency and the American NASA have carried out in the last decades. In this chapter i will describe the main discoveries, taking into account the different regions where cometary bodies have been identified (Fig. 3.4).

The first billion years of the solar system were convulsive and marked by large collisions, and ultimately by the scattering of icy bodies under the effects of the migration of the giant planets (Gomes et al., 2005). Models that can now reproduce the mass and distribution of the planets suggest that about 3,800 Ma ago a migration of Jupiter and Saturn into the inner solar system took place under gravitational interaction with an asteroid belt that formed several orders of magnitude denser than in the present time. The outer region of this belt was very massive and made up of transitional objects in the style of the hydrated planetoid Ceres. The remaining transitional objects are called asteroid belt comets and are objects that may possess rocky nuclei resembling carbonaceous asteroids. The gravitational scattering processes induced by the giants were so intense that it is thought that we might even find such objects in the Oort Cloud. These objects would have their maximum distance from the Sun (aphelion) at tens or even hundreds of thousands of times the mean Earth–Sun distance, and their orbits could be stable but highly eccentric. Because of this, and as a direct consequence of Kepler's area law, the high relative velocities at which these objects would collide with our planet would constitute one of the greatest latent dangers to life on Earth in time periods of hundreds of millions of years.

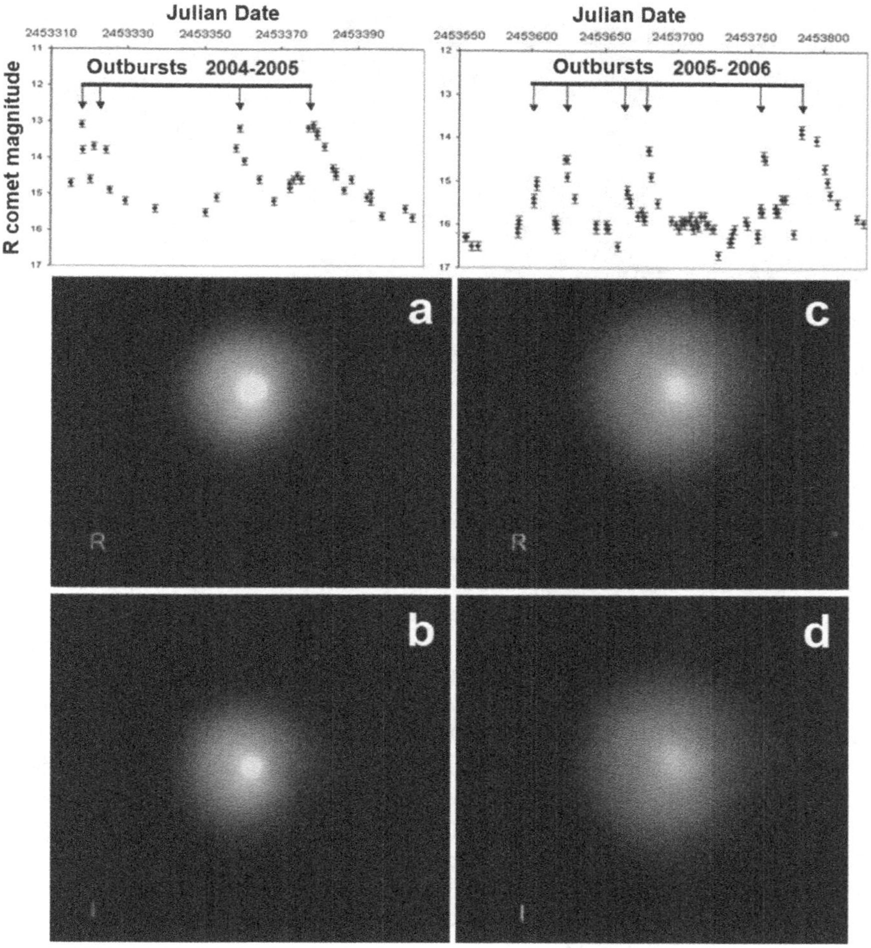

Fig. 3.3 Example of photometric tracking of the brightness magnitude of Comet 29P/Schwassmann–Wachmann 1. Each point in the graphs corresponds to an observation using a CCD camera as photometric sensor. On the upper axis the time is expressed in Julian days. Below are images of this comet obtained by the author on successive nights to appreciate the expansion. The IAC-80 telescope was used using R and I filters. **a–b** February 5.09, 2010; and **c–d** February 6.08, 2010

Thus, while some comets are essentially fragile objects, having preserved their primordial properties and having remained in distant regions of the solar system, others may pose an enormous potential danger in the event of a direct impact with the Earth.

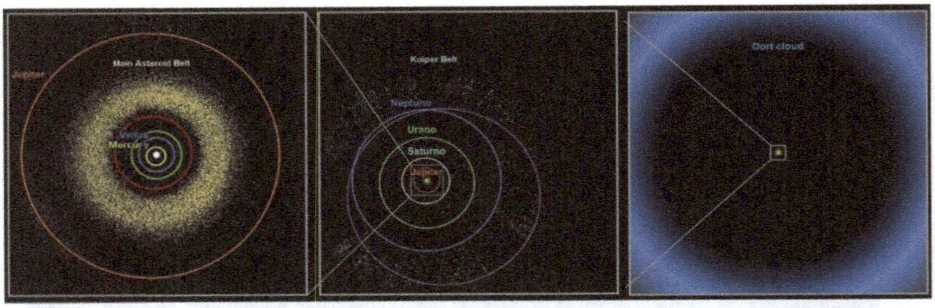

Fig. 3.4 The three regions of the solar system where primitive objects have been stored. Successive scale changes are necessary to reach the outer regions of the Kuiper Belt and the Oort Cloud (adapted image: NASA/JPL-Caltech/R. Hurt)

3.1 Impact with Comets

Several authors have invoked in the scientific literature the existence of cometary showers which, if they occur again in our times, could threaten our civilisation (Hut et al., 1987). Regardless of the fact that a long- or short-period comet with an eccentric orbit that is close to the Earth's orbit could disintegrate and generate a swarm of fragments large enough to be of concern, there are also plausible mechanisms capable of producing the dispersion of icy objects towards the terrestrial planets. The effects of galactic gravitational tides or of the solar system passing alongside neighbouring stars was modelled by Matese et al. (1995), who derived a plausible periodicity of between 30 and 35 Ma. Indeed, Shoemaker (1998) suggested that there is evidence for such a periodicity in the ages of the impact craters and impactites we know of, but also in the amounts of ^{3}He (of interplanetary origin and presumably incorporated into the projectiles that hit us) accumulated in the stratigraphic record. Farley et al. (1998) described geological evidence for a late Eocene cometary shower based not only on the existence of ^{3}He, but also of significant peaks in the abundance of iridium (Ir) measured in 35 and 36 Ma strata.

Other studies attempted to quantify in statistical terms the possible contribution of comets. Fernández and Sosa (2012) completed a detailed study based on the historical compilation of comets observed since 1900. Based on these astronomical observations over more than a century they were able to infer interesting details about the size distribution of long-period comets. They found that long-period comets have a size distribution that suggests they tend to fragment as they pass through perihelion, creating several daughter comets. It appears that the size distribution indicates that comets smaller than 500 m are extremely rare, despite increased tracking of NEOs over the past decades. These dimensions are therefore indicative of the minimum size a comet must possess for its nucleus to remain active and visible. They also found that about 30% of these long-period comets are new and that, among them, about half come from the outer Oort Cloud and half from the inner Cloud. Our experience observing comets in the inner planetary system indicates that their decay is a plausible outcome. This sudden process could generate dangerous projectiles since their orbits and high eccentricities give them much higher impact energies compared with projectiles arriving from the main belt.

Pioneering Missions to Comets

The fascination to know more details about Comet 1P/Halley led different space agencies to design a whole armada of spacecraft to closely explore the nucleus of the comet, usually hidden in the coma of sublimated gas in its interior. The plethora of results obtained by these probes that passed through the comet's coma in 1986 was a first breakthrough in the enigmatic origin of these objects, although many other questions were raised. In recent decades, the scientific community has been accumulating evidence that these are primitive objects which preserve within them the keys to the physico-chemical processes that took place during the origin of the solar system. To corroborate these ideas, NASA's *Stardust* mission in 2006 recovered, for the first time, particles from a comet, catalogued as 81P/Wild 2; these are in fact the first solid samples recovered in situ from a solar system object other than the Moon (Brownlee et al., 2006). The studies of that probe revealed that Comet 81P/Wild 2 has a diameter of 4.5 km and is mostly composed of a mixture of ices, organic matter and mineral grains, whose typical size is about 10 μm (a micrometre is one thousandth of a millimetre) (Fig. 3.5).

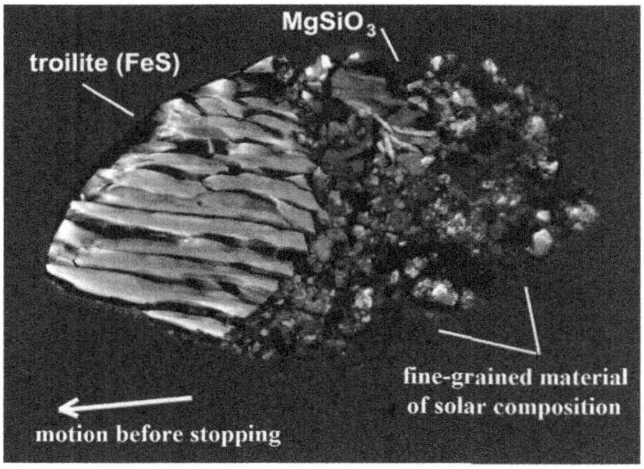

Fig. 3.5 One of the particles from Comet 81P/Wild 2 recovered by the Stardust probe wrapped in aerogel. The presence of troilite (FeS) at the front of the Febo particle shielded the effect of the thermal front. This preserved the enstatite ($MgSiO_3$) and the thin material (with CI carbonaceous chondrite-like composition) behind it, which would otherwise have been vaporised. An arrow indicates the direction of the particle's motion before it was stopped in that position For more details see G. Matrajt et al. 2008 (Stardust PET/NASA)

Another important NASA mission was the *Deep Impact* mission that launched a 370 kg metallic projectile on comet 9P/Tempel 1. This impact generated a crater and a significant emission of dust and gas in the resulting impact plume. However, such an experiment had a rather small effect on the comet, exemplifying the enormous intrinsic difficulty behind attempting the destruction or deflection of such a comet. It should be remembered that this comet has an oval shape with axes of 7.6 × 4.9 km. Perhaps one of the most unexpected results of NASA's *Deep Impact* mission was that Comet 9P/Tempel 1 produced dust particles of a crystalline nature. The crystalline silicates formed in close proximity to the Sun and thus contrasted with the supposedly pristine nature representative of regions far from the protoplanetary disc in which it would have formed. These data, obtained from the remote study of dust properties, needed to be corroborated in other comets. This was the role of the Stardust mission, which I will discuss later.

The icy objects in the Oort Cloud and the Trans-Neptunian belt are the most abundant in the solar system. It is estimated that there must be billions of them and that they are basically composed of ice and small mineral particles. Because of their relatively small size and high percentage of ices and organic matter, it is estimated that they have been weakly thermally processed as if they were rocky bodies that incorporated significant proportions of radioactive elements that would eventually melt the interiors of those asteroids of several hundred kilometres. These objects have spent most of their time orbiting a very cold region far from the Sun and their materials are the most primitive we can find.

The precursors of medium- and short-period comets are bodies that have escaped from the Cloud and the Belt and enter the region close to the Sun by gravitational perturbations. From the *Oort Cloud*, comets arrive in long-period orbits with orbits reaching orbital periods of tens or hundreds of thousands of years. Given their distance, their intrinsic smallness and their minimal ability to reflect the faint light they receive from the Sun, Oort Cloud objects can only be detected if they transform into comets during their transits through the inner solar system. It is therefore not surprising that no comets have been observed in the Oort Cloud. Despite being a theoretically postulated region, it explains the continuous annual discovery of dozens of comets from these distant regions at the limit of solar influence.

3.2 The Stardust Mission and the 81p/Wild 2 Materials

Comets are fragile bodies, and collecting their forming materials is more challenging than we would initially think. For that reason, the Stardust NASA mission, the first sample-return mission to bring materials from a comet (and from any body other than

Fig. 3.6 Artist's impression of the deployed honeycomb collector during the Stardust probe's path through the coma of Comet 81P/Wild 2 (Stardust/NASA)

the Moon), used a really smart technique using a material called aerogel. Although the first aerogel was created around the 1930s, its enormous potential for capturing cosmic and cometary dust was developed within the planetary science community. Particles from Comet 81P/Wild 2 were captured by the Stardust probe (Fig. 3.6) thanks to an ingenious honeycomb collector mechanism in whose cells were housed blocks of aerogel of identical size and properties. The aerogel had a density that varied gradually from $\rho < 0.01$ g/cm^3 at the surface to 0.05 g/cm^3 at about 3 cm deep. This characteristic is important, as it was intended to allow gradual capture while minimising the duration of the thermal pulse that could chemically alter the particles. Since the particles reached the collector at a relative velocity of 6.1 km/s, the effects of the alteration can be quantified (Trigo-Rodriguez et al., 2008). While the kinetic energy of the incident particles may be sufficient to evaporate them on direct impact against a solid surface, their capture in a low-density aerogel allows only the more volatile phases to be evaporated, and the material at the shock front acts as a thermal shield for the material behind it. The range of heating and deceleration is defined by the amount of momentum and energy available at each instant, as we will see in the next section. Evidently, these magnitudes change rapidly during capture, not only due to the progressive deceleration of the projectile in the aerogel but also due to the loss of mass experienced as a consequence of heating and fragmentation. The largest aggregates penetrating the aerogel had volatile phases that, when the particles fragmented shortly after the aerogel capture, contributed to produce bulbous tracks (see, for example, Fig. 3.7). I led the study of these capture processes in Stardust, so we were able to infer the heating experienced during the capture of these materials (Trigo-Rodríguez et al., 2008c).

Comet 81P/Wild 2 has a diameter of 4.5 km and is mostly composed of a mixture of ices (mainly water), organic matter and mineral grains that are mainly in the form of fine dust of a typical size of about 12 μm (Brownlee et al., 2006). When, after its encounter with Jupiter, this comet moved into an orbit closer to the Sun,

Fig. 3.7 Some of the tunnels dug by Comet 81P/Wild 2 particles during their capture in the aerogel. The first three are type B while the last one (T22) is type A. The particles penetrated the aerogel from above (Stardust PET/NASA)

the increased irradiation resulted in a higher surface temperature that induced the increasing sublimation of the water ice. This process promotes the emission of gas, which in turn drives the dust particles retained in its porous structure. This mechanism, first described by Fred L. Whipple, acts by reducing the mass of the comet by several tonnes per second and the sublimated gas carries the dust particles away. When the Stardust probe passed through the inner regions of this comet's coma and deployed its collector, thousands of particles reached it. In this process of particle deceleration in the aerogel, the captured particles excavate cavities that I will call here '*tracks*', with typical sizes similar to those of the captured projectile. Given the difficulty of modelling the entry of a solid at high velocity into a low-density medium, much of what is known about the relationships between the dimensions of such tunnels and the properties of the projectiles that excavated them has been obtained in laboratory experiments. Different laboratories around the world have specialised in such experiments in which particles of defined composition and diameters are fired at different aerogels in order to study the strength of such materials and the effects suffered by the particles. These materials are highly insulating and have all kinds of applications, both in our laboratories and in space.

The chemical and isotopic heritage preserved in the interior of Comet 81P/Wild 2 may also reveal secrets about the environment where this comet was born (Brownlee et al., 2006). Indeed, it has been surprising to discover that the materials themselves are not very different from those that form carbonaceous chondrites. The asteroids represented by carbonaceous chondrites have been able to retain up to 10% water (by mass) and 4% organic matter, although unlike Comet 81P/Wild 2 they must have undergone significant compaction processes due to collisions with other objects (Blum et al., 2006; Trigo-Rodriguez and Blum, 2009). The compaction and the heating released by the shock waves in these collisions could indeed be the origin of these meteorites having undergone extensive aqueous alteration processes (Trigo-Rodríguez et al., 2006). Precisely this topic was the focus of my studies on carbonaceous chondrites at the University of California Los Angeles (UCLA), where we found that most of these objects have hydrated mineral phases (Rubin et al., 2007). It has been a great surprise to many researchers that the mineral grains recovered from Comet 81P/Wild 2 were not hydrated, possibly showing that water never soaked these materials as it did in the progenitor asteroids of CM-type carbonaceous chondrites (Trigo-Rodriguez et al., 2006, 2007).

One of the most fascinating surprises of the preliminary examination stage of the *Stardust* samples was to find crystalline silicates formed at high temperatures in the vicinity of the Sun in the aerogel (Brownlee et al., 2006). Examples are mineral grains of forsterite (Mg_2SiO_4), enstatite ($MgSiO_3$) or even an inclusion similar to refractory oxides known as Ca–Al-rich inclusion (*CAIs*). It was not thought that such materials could be so common in a comet formed in the remote Kuiper Belt (Fig. 3.4). For these materials, formed at high temperature in regions of the solar nebula closer to the Sun, to be incorporated into this comet formed beyond Neptune has required a significant radial flow of material that must be explained by a transport of the materials through turbulent regions. This changes the preconceived idea that the materials contained in comets had formed at great distances from the Sun. Possibly an intense solar wind coupled with turbulent transport mechanisms in the protoplanetary disc were able to sweep such refractory minerals to distant regions where they were incorporated into the comets.

The notion that comets were primitive objects with a fascinating and fundamental chemistry in the enrichment in water and organic matter of the early Earth was predicted in 1961 by Joan Oró (1923–2004). Ongoing studies of particles recovered from comet 81P/Wild 2 are beginning to bear great fruit and support the idea that many comets have undergone minimal alteration since their primordial aggregation (Brownlee et al., 2006). NASA is entrusting samples to consolidated groups equipped with highly sophisticated instruments. The *Stardust* mission marks the beginning of the study of other planetary bodies beyond the Moon. We will have to make a major effort if we wish to contribute actively to the study of primitive materials that hold the keys to the origin of our solar system.

3.3 The Rosetta Mission and Comet 67P/Churyumov–Gerasimenko

The interplanetary probe Rosetta, a complex device created by the European Space Agency, several European agencies and NASA, which is already orbiting its final target, the Comet 67P/Churyumov–Gerasimenko, deployed its lander Philae to approach the surface of this primitive object. Of the 25 scientific experiments contained in the Rosetta probe, about ten were part of the state-of-the-art instrumentation that Philae planned to use for physico-chemical analysis of the comet's surface, subsurface and internal structure. The Rosetta mission has made enormous progress in our knowledge of comets and has made it possible to map and observe cometary activity as never before. The result of this work is thousands of iconic images of the grandeur and complex structure of these comets, which, in this case, were shown to have two lobes (Fig. 3.8). The scientific work of the Rosetta mission (ESA) has given rise to dozens of research articles and would merit a separate book. Here, for the sake of simplicity and for the sake of information, I will briefly describe some of its instruments, quoting their acronyms.

The *Philae* lander with a total mass of about 98 kg was equipped with legs and a grappling mechanism that ultimately failed to complete the risky manoeuvre. It seems that the comet was so fragile that the lander passed through some of the its formative material and got stuck in another area after bouncing several times (see Fig. 3.9). *Philae* was to perform a series of unique experiments in a relatively short period of time, but some of them could not be carried out. It still took the closest image we have of a comet and performed valuable experiments for only four days because its primary battery was not rechargeable and had a capacity of around 1,200 W/h.

Philae was equipped with the CIVA and ROLIS cameras, both panoramic and low-detail cameras, to capture high-resolution images of the surface activity and the material that forms the comet. In addition, it was equipped with the SESAME

Fig. 3.8 The two lobes of Comet 67P captured by the Rosetta probe, showing the jets of gas and particles being launched from its myriad active zones (ESA/Rosetta/NAVCAM)

Fig. 3.9 The Agilkia region chosen for Philae's landing, showing the final landing site (ESA/Rosetta/MPS/MPS for OSIRIS Team MPS/UPD/LAM/IAA/SSO/INTA/UPM/DASP/IDA)

instrument, an acoustic probe that could detect the impact of the dust particles that are ejected from the surface as a result of gas sublimation.

Further, the CONSERT experiment was able to receive and transmit radio signals to and from the Rosetta orbiter, allowing it to probe the structure and constitution of the cometary nucleus on a large scale. In short, the plan was to get a glimpse of its internal structure to answer fascinating questions about the formation of these icy stars.

PTOLEMY is a mass spectrometer that can delve into the isotopic composition of hydrogen, carbon, nitrogen, oxygen, sulphur and other stable isotopes. Morse et al. (2015) were able to use this instrument to show that there are significant compositional differences in different sublimated regions of Comet 67P/Churyumov–Gerasimenko. They focused their study on the CO/CO_2 ratio of the sublimated ices component in the surface environment of the Abydos region, a value substantially lower than that measured by another instrument (ROSINA) in the comet's coma. Such observations show that comets have highly heterogeneous compositions.

In addition, COSAC is a gas chromatograph and time-of-flight mass spectrometer (TOF–MS) that allows the identification of the molecular composition of the material, in particular of the organic components. With these instruments, the chemical and isotopic composition of the formatives was further investigated. This is particularly important because, with the exception of the recovery of samples from Comet 81P/Wild 2 by NASA's Stardust mission, the material from a comet itself has never been studied before, let alone in situ as Philae did.

The CONSERT experiment revealed that the minimum content of carbonaceous material to explain the signal was 75% by volume. This suggests that comets represent a huge reservoir of carbon (Kofman et al., 2015). The comet was also not heterogeneous but possessed a fine aggregate structure on a few wavelength scales. The mean permittivity obtained with this instrument was 1.27, indicating that the region studied had a volumetric dust-to-ice ratio between 0.4 and 2.6, as well as a porosity of 75–85%. These results would be consistent with a fine-grained matrix comparable to that of carbonaceous chondrites.

In the context of the impacts of these objects on Earth, the lessons of the Rosetta mission are undoubtedly enormous. To briefly exemplify the grandeur of the challenge posed by a medium-sized comet like 67P/Churyumov–Gerasimenko, it must be remembered that the physical properties of such an object (low consistency and density) make a deflection by direct impact impossible. Knowing the structure of the object, a dedicated mission could consider the use of some kind of projectile in a particularly fragile region—an example could be the neck region joining the two lobes, which was also an ice-rich and active region. In fact, we often observe comets fragmenting into small pieces and finally vanishing. Obviously, allowing such an enormous object to impact would result in a catastrophe on a planetary scale and with consequences that are difficult to pinpoint.

Let's consider the piled and chemically heterogeneous nature of Comet 67P/Churyumov–Gerasimenko with its estimated volume of 18.7 km^3 that corresponds to a total mass of ten billion tonnes. A high-speed impact by a comet having such dimensions would be a dangerous chemical cocktail for life on Earth. Such an impact would not only involve a huge explosive process due to the phase transition of the low consistency materials under the overpressure collapse of the object, but would also cause an injection of other reactive chemical elements with direct consequences for the environment in the short and medium term: S, P, Mg, Na, etc.... The chemical effects of such an impact plume for the entire biosphere could be certainly lethal (McLaren and Goodfellow, 1990).

Obviously, the problem of discovering new comets in sufficient time to be able to study them and develop appropriate mitigation strategies is of the first magnitude. They are extremely dark and poorly reflective objects that are very difficult to see at great distances. If we want to discover these new kilometre-scale comets approaching the inner solar system from distant corners of our planetary system, we will have to be ingenious. To detect them at the distance of Jupiter, a rough estimate is that we would have to deploy 30 10-m aperture telescopes on Earth, or even several tens of telescopes in stable regions around Jupiter, such as the so-called Lagrange points. These projects seem extravagant and far from our real possibilities.

3.4 The Trans-Neptunian Belt

Just as we speak of the asteroid belt to refer to the storehouse of objects that have been preserved between Mars and Jupiter, we can also speak of another region that harbours a very significant number of objects much smaller than the planets. These bodies are mostly rich in ices and organic matter, although, given their enormous distances from the Sun, their behaviour is generally dull and their identification highly elusive.

Trans-Neptunian objects (TNOs) are not only considered the 'parents' of these so-called centaurs but also of short-period (<20 years) comets (Jewitt, 2008). Much progress has been made in the knowledge of these objects in the last decade. Among the most notable events related to this population, the discovery in 2005 of three TNOs of similar size to Pluto should be highlighted: Eris, Makemake and 2003 EL61—the latter was discovered at the Sierra Nevada Observatory by the group of the Instituto de Astrofísica de Andalucía (CSIC) led by Dr. José Luis Ortiz, who has made important contributions to a better understanding of the dimensions and physical properties of TNOs (Sicardy et al., 2011; Ortiz et al., 2012). Table 3.2 compiles the dimensions of the largest Trans-Neptunian objects known to date.

The fragmentation in cascade of these fragile objects in the Trans-Neptunian region means that, in their progressive loss of potential energy, they end up crossing

Table 3.2 Major trans-neptunian objects with a diameter greater than 1,000 km

Object	Diameter (km)
(136,199) Eris	2,326
(134,340) Pluto	2,340
(90,377) Sedna	1,600
(136,472) Makemake	~1,500
(136,108) 2002 EL61	1,270
Charon (Pluto's satellite)	1,210
(84,522) 2002 TC302	1,150
(50,000) Quaoar	1,140

the orbits of the giant planets. These dynamic paths may involve their progressive dispersal into the inner solar system or towards an apotheotic end on a collision course towards the giant planets (Jewitt, 2008). Some of these objects are in fact absorbed by the gravitational fields of the giant planets and produce huge fireballs (Hueso et al., 2018). We have evidence of such close encounters in the observation of large fireballs in Jupiter's atmosphere, for example. Likewise, the capture and subsequent fall on the planet Jupiter of Comet Shoemaker–Levy 9 was a good example of the protective role played for life on Earth by that giant planet. Long-period, low-inclination comets (i.e., in an Earth-like orbital plane) have a high probability of encountering the massive Jupiter, which will simply exert a gravitational influence that will alter their course and fate. We now know that such encounters frequently produce fragmentations and tend to shorten the orbits of comets. The fragmentation of these comets due to internal stresses, caused by the gravitational tidal effect in the vicinity of the planet, means that the centaurs tend to lose size and, therefore, destructive potential. Objects that decay in their orbit below that of Jupiter will eventually cross the asteroid belt and may eventually interact with the rocky planets. The evolution of short-period comets and their impact hazard will be the focus of the next chapter.

The size distribution of comets of the Jupiter family seems consistent with that which would emanate from these collisional processes between TNOs (Tancredi et al., 2006). On the other hand, that study of the fractions of the surface of these objects where cometary activity is maintained—i.e. where the sublimation of ices and the consequent emission of dust occurs—found it is minimal, generally less than 1%, which led them to suggest that comets of this family are already evolved and eroded objects that move towards stages of very low activity or even dormancy.

Questions for Chapter 3

Q.1. Comets share amazing physical properties, exhibiting extremely low bulk densities and tensile strengths. They are among the most fragile large bodies in space. Recent studies by Rosetta spacecraft confirmed our estimations using the disruptive strength of cometary meteoroids in the atmosphere (Trigo-Rodríguez & Blum, 2009). Mould sands (e.g., quartz sands under laboratory conditions) display wet tensile strengths of about 0.5 N/cm^2. Just to give you a rough idea, a comet like 67P/Churyumov–Gerasimenko exhibits a tensile strength of about 0.1 $^N/cm^2$. Search for other materials exhibiting similar strength values, and try to explain how to build a km-sized comet using such type of fragile materials. Do you know materials on Earth of similar tensile strength? [Solution: We really don't know such fragile materials on Earth with the exception of unstable materials. Who has not played, for example, on the beach with extremely soft wet sand?].

Q.2. Taking into account the main compositional results obtained by Stardust (NASA) and Rosetta (ESA) missions respectively to 81P/Wild 2 and 67P/Churyumov–Gerasimenko, try to explain the possible implications of the impact of a comet with the atmosphere of the Earth. Try to enumerate the different key processes of a fragile body, rich in volatile species, penetrating in the atmosphere at hypervelocity. [Solution: The impact plume formed after the disruption of a comet in a planetary atmosphere might be at some extent a kind of lethal chemical bomb. We should take into account that comets contain significant amounts of sodium, phosphorus, sulphur, etc. The combination of these elements in an oxidizing environment could produce molecular compounds with lethal consequences for living species if reaching the ground and the oceans in significant amounts.]

Q.3. Try to make a comprehensive list of large TNOs discovered so far (Table 3.2 just cites the largest ones). Compare the diameters of the largest with that of Pluto. Do you think it was appropriate to retire the status of planet from Pluto? Give your own arguments, and compare with the definition of planet. [Solution: In the Trans-Neptunian region we cannot claim that a planet has cleaned its vicinity of small bodies, as there are many objects of a similar size to Pluto.]

Online Resurces

The International Comet Quarterly (ICQ) Information Website: http://www.icq.eps.harvard.edu/

Recent comet magnitude estimates: http://icq.eps.harvard.edu/CometMags.html

References

Boehnhardt, H., N. Rainer, K. Birkle and G. Schwehm (1999) The nuclei of comets 26P/Grigg–Skjellerup and 73P/Schwassmann–Wachmann 3. *Astronomy and Astrophysics* 341, 912–917.

Blum J., R. Schräpler, B.J.R. Davidson and J.M. Trigo-Rodríguez (2006) The physics of protoplanetesimal dust agglomerates. I. Mechanical properties and relations to primitive bodies in the solar system, *Astrophysical Journal* 652, 1768-1781.

Brownlee et al., 2006.Brownlee D. et al. (2006) Comet 81P/Wild 2 under a microscope. *Science* 314, 1711–1716.

Farley K.A., A. Montanari, E.M. Shoemaker and C.S. Shoemaker (1998) Geochemical evidence for a comet shower in the late Eocene. *Science* 280, 1250–1253.

Fernández, J. A. and A. Sosa (2012) Magnitude and size distribution of long-period comets in Earth-crossing or approaching orbits. *Monthly Notices of the Royal Astronomical Society* 423, 1674–1690.

Gomes R., H.F. Levison, K. Tsiganis and A. Morbidelli (2005) Origin of the cataclysmic Late Heavy Bombardment period of the terrestrial planets. *Nature* 435, 466–469.

Hut P., W. Alvarez, W.P. Elder, T. Hansen, E.G. Kauffman, G. Keller, E.M. Shoemaker, and P.R. Weissman (1987) Comet showers as a cause of mass extinctions. *Nature* 329, pp. 118–126

Hueso, R. et al. (2018) Small impacts on the giant planet Jupiter. *Astronomy & Astrophysics* 617, A68, 13 pp.

Jewitt D. (2004) Looking through the HIPPO: Nucleus and dust in Comet 2P/Encke. *The Astronomical Journal* 126, 3061–3069.

Jewitt D. (2008) Kuiper Belt and Comets: An observational perspective. In *Trans-Neptunian Objects and Comets, Swiss Society for Astrophysics and Astronomy*, Springer-Verlag, Heidelberg, Germany, pp. 1–78.

Kofman, W. et al. (2015) Properties of the 67P/Churyumov–Gerasimenko interior revealed by CONSERT radar. *Science* 349, Issue 6247, aab0639.

Matese J.J., P.G. Whitman, K.A. Innanen, and M.J. Valtonen (1995) Periodic modulation of the Oort cloud comet flux by the adiabatically changing galactic tide. *Icarus* 116, pp. 255-268.

Matrajt G., M. Ito, S. Wirick, et al. (2008) Carbon investigation of two Stardust particles: A TEM, NanoSIMS, and XANES study. *Meteoritics & Planetary Science* 43, 315–334.

McLaren and Goodfellow, 1990.McLaren D.J. and W.D. Goodfellow (1990) Geological and biological consequences of giant impacts. *Annual Review of Earth and Planetary Sciences* 18, 123–171.

Morse, A., O. Mousis, S. Sheridan, G. Morgan, D. Andrews, S. Barber, and I. Wright (2015) Low CO/CO2 ratios of comet 67P measured at the Abydos landing site by the Ptolemy mass spectrometer. *Astronomy & Astrophysics* 583, id.A42, 4 pp.

Ortiz J.L. et al. (2012) Albedo and atmospheric constraints of dwarf planet Makemake from a stellar occultation, *Nature* 491, 566–569.

Rubin A.E., J.M. Trigo-Rodríguez, H. Huber, and J. T. Wasson (2007) Progressive aqueous alteration of CM carbonaceous chondrites. *Geochimica et Cosmochimica Acta* 71, 2361-2382.

Schambeau, C., Y. Fernández, C. Lisse, N. Samarasinha and L. Woodney (2015). A new analysis of Spitzer observations of Comet 29P/Schwassmann–Wachmann 1. *Icarus* 260, 60–72.

Shoemaker E.M. (1998) Impact cratering through geologic time. *Journal of the Royal Astronomical Society of Canada* 92, 297–309.

Sicardy B. et al. (2011) A Pluto-like radius and a high albedo for the dwarf planet Eris from an occultation. *Nature* 478, 493–496.

Tancredi, G., et al. (2006) Nuclear magnitudes and the size distribution of Jupiter family comets. *Icarus* 182, 527–549.

Trigo-Rodríguez J.M. and J. Blum (2009) Tensile strength as an indicator of the degree of primitiveness of undifferentiated bodies. *Planetary and Space Science* 57, 243–249.

Trigo-Rodríguez J.M., A. Rubin and J.T. Wasson (2006) Non-nebular origin of dark mantles around chondrules and inclusions in CM chondrites. *Geochimica et Cosmochimica Acta* 70, 1271–1290.

Trigo-Rodríguez J.M. et al. (2007) Comet 17P/Holmes. *Central Bureau Electronic Telegrams*, No. 1118, #1 (D.W.E. Green ed.).

Trigo-Rodríguez J.M., E. García-Melendo, B.J.R. Davidsson, A. Sánchez, D. Rodríguez, J. Lacruz, J.A. de los Reyes and S. Pastor (2008a) Outburst activity in comets: I. Continuous monitoring of Comet 29P/Schwassmann–Wachmann 1. *Astronomy and Astrophysics* 485, 599–606.

Trigo-Rodriguez, J. M., B. Davidsson, P. Montanes-Rodriguez, A. Sànchez and B. Troughton (2008b) All-sky cameras detection and telescope follow-up of the 17P/Holmes Outburst. *39th Lunar and Planetary Science Conference, League City, Texas.* LPI Contribution No. 1391, p.1627.

Trigo-Rodríguez et al., 2008c.Trigo-Rodríguez J.M., G. Domínguez, M. J. Burchell, F. Hörz and J. Llorca (2008c) Bulbous tracks arising from hypervelocity capture in aerogel. *Meteoritics and Planetary Science* 43, 23–40.

Trigo-Rodríguez J.M., D.A. García-Hernández, A. Sánchez, J. Lacruz, B.J.R. Davidsson, D. Rodríguez, S. Pastor and J.A. de Los Reyes (2010) Outburst activity in comets – II. A multi-band photometric monitoring of Comet 29P/Schwassmann–Wachmann 1. *Monthly Notices of the Royal Astronomical Society* 409, 1682–1690.

Chapter 4
The Risk Associated with Short-Period Comets and Its Origin

When talking about the danger of cometary impact, a distinction must be made between the rare chance encounters with long-period comets and the more frequent encounters with evolved short-period comets. In fact, the comets most likely to stalk our planet are the so-called NEC (Near-Earth Comets) of which we currently know of about a hundred. In fact, it may surprise the reader that most of these comets do not behave like comets and could even be pigeonholed as asteroids. This is because the evolution of short-period comets inexorably involves processes of fragmentation and progressive weathering as a consequence of their multiple passes to the Sun (Fig. 4.1). This is particularly true for short-period comets that are driven to multiple approaches to the Sun and wear out their content of volatile compounds (mostly ices). Some of these objects have perihelia—points in their orbits of closest approach to the Sun—that are so narrow they can even lead to the weathering of other mineral compounds due to their proximity to surface temperatures of hundreds of degrees Celsius.

In fact, based on their orbits, it is possible to identify these comet fragments, which are sometimes pieces of large comets that disintegrated hundreds or thousands of years ago. These remnants clearly point to the fact that multiple comets have disintegrated over the past hundreds of thousands of years in their transits into the inner solar system, particularly in their encounters with Jupiter. Possibly these catastrophic fragmentations arise from the inherently fragile nature of comets (Brownlee et al., 2006), as I outlined in the previous chapter. And that brings us to one of the greatest challenges facing humanity: to properly catalogue and track those very dark (low albedo, as we astronomers say) and highly eccentric objects that are still hidden in the population of potentially dangerous bodies. Most of these bodies could be hidden as dark asteroids that are probably not active anymore, being covered by a layer of rubble that avoids the sublimation of the remaining ices. There is possibly no better example than the extinct Comet 2015 TB145 as evidence of the disruption of a highly eccentric body, and this particular case is one I will discuss in detail at the end of this chapter.

J. M. Trigo-Rodríguez, *Asteroid Impact Risk*, Impact Studies,
https://doi.org/10.1007/978-3-030-95124-5_4

Fig. 4.1 Comet 2P/Encke in an image made by Damian A. Peach on 12 October 2013 (courtesy of the author)

4.1 The Fragile, Dark and Elusive Nature of Comets

When thinking of impact hazards, it is impossible to ignore the comets. Often it is invoked that a strike by comets is infrequent, because we consider long-period comets in high inclination orbits—but what about evolved comets? Many comets, being fragments of much larger objects, have been circling our planet in very short-period orbits. Some of them seem to be not as fragile as we thought. Even when every few years they pass close enough to the Sun to heat up their surfaces significantly, they persist in threatening our planet. In addition, they have been encountering rocky bodies that have impacted on their surfaces over the eons, causing these short-period comets to become less active—i.e. the percentage of active sublimation regions is progressively reduced in a natural way. As a consequence, short-period comets usually become extinct. They end up covered with rocky regolith due to the landscaping of the surface effected by the rocky materials expelled by the cometary activity itself, as well as by the countless impacts with other objects in the debris-ridden regions. Consequently, the materials that form these objects are thought to be the product of multiple impacts that contribute to produce brecciated materials with peculiar reflective responses.

It is precisely our search for good prototypes of the materials that make up the nature of comets that has allowed us to discover that certain ungrouped carbonaceous chondrites—i.e. those not belonging to the main groups of this type of chondrite—have reflection spectra similar to those of Comet 2P/Encke (Tanbakouei et al., 2020).

The type of materials formed are called breccias. They have different rock types, technically called lithologies, which generally point to a complex formation, produced by impact with bodies formed in different regions of the solar system.

The physical properties of these materials are a challenge in themselves, as they can vary at different scales given their heterogeneous nature, sculpted by countless processes: heat, aqueous alteration, and so on. However, they are among the least reflective materials we know of and are often extremely fragile (Trigo-Rodriguez et al., 2014). Needless to say, this means that they could well pass us by unnoticed, as I explain at the end of this chapter. When materials are so heterogeneous and have appreciable porosity, the thermal conductivity is so low that regions rich in volatile substances can be found just a few centimetres below. Despite of it, if the object reaches an orbit with a close perihelion the object can fracture by a gravitational tidal effect, so the volatile species can be irradiated to suffer a change of phase promoting subsequent aqueous alteration reactions inside of it. Once that process occurs I envision that the action of water corrodes internally the body as the interaction with some reactive minerals takes place. In fact, we discovered that in such a phase Fischer–Tropsch reactions can take place producing complex organic molecules (Rotelli et al., 2016; Cabedo et al., 2021). Over time these crumble objects become extremely weak and tend to disintegrate as we usually observe in comets. In fact, the extremely rare CI chondrites are composed by extensive aqueous alteration materials that could have experienced similar processes (Trigo-Rodríguez et al, 2019). As a result, developing an appropriate strategy to mitigate a potentially dangerous encounter with such objects could be a daunting challenge.

4.2 Impacts, Atmospheric Bursts and Their Markers

If we were to compare the danger of impact with a mythological analogy, we could choose that of the Hydra, the mythical multi-headed animal. We will see why in this chapter. There is no doubt that we live quite oblivious to the danger that looms over our existence on Earth. Perhaps it is because we have not yet learned to recognise the dangers, ominous in their magnitude and diversity. We know that the atmosphere is a good natural shield for asteroids from a few tens to a hundred metres in diameter. This is because impact craters smaller than a kilometre are rare and, on average, the projectiles that excavate them are an order of magnitude the size of the crater (Bland and Artemieva, 2003). This indicates that objects with a diameter of less than a hundred metres tend to fragment efficiently and rarely touch the ground with the potential to excavate a crater. This evidence could give us some comfort, given that these projectiles are the ones that could most frequently affect us on a human timescale. However, before we stop worrying too much, we should be aware that cratering is by no means the only way to cause a catastrophe. I will reason this out below, with examples from our relatively recent history.

Both chondritic asteroids and comets, the latter being much more fragile, may disintegrate before reaching the ground, with no opportunity to excavate a crater, but with a destructive capacity that is not negligible. As these objects deepen in the atmosphere at hypervelocity, they feel the shock wave and the increasing hydrodynamic pressure, which increases proportionally to the density in each stratum and is multiplied by the velocity of the object squared, so they tend to compress until they reach the fracture point. After fragmentation, pulverised material is generated, which is rapidly evaporated. This results in an explosion (or several explosions) in which the fragments are dispersed under the vapour pressure experienced. The larger fragments of the projectile retain some of the kinetic momentum and continue their flight, but the mass of the object is drastically reduced, and thus the danger. Many fireball observers have reported the existence of flares in the final sections of activity and can even hear the explosions due to the shock waves generated.

At the end of the luminous section, the result is a huge bolide consisting of dust and gas at a few thousand degrees Celsius. The mass of rocky objects with a diameter of about 50 m could already be sufficient, depending also on the geometry of the trajectory, for such a fireball phase to reach the ground (Trigo-Rodríguez et al., 2021). We find in the recent literature evidence for events whose nature was difficult to explain until recently. There are certain layers in the stratigraphic legacy that are rich in organic matter with clear markers of extraterrestrial origin. There are also certain regions where we find tektites: glasses produced by the sudden heating of rocks on the Earth's surface without associated impact craters. Are there phenomena that can induce an abrupt heating of the Earth's surface without the need to excavate a crater? The answer is yes, and such phenomena are known as low-level atmospheric outbursts. Perhaps, on a small scale, the best example was the Tunguska event over the Siberian region of the same name in 1908. I will describe that fascinating case later, but now let us look at the origin of these phenomena and the evidence that has been collected.

It turns out that during the sudden entry through the atmosphere of objects of cometary nature it is to be expected that they will be deformed and fragmented by aerodynamic forces. As I noted earlier, when they penetrate the atmosphere the hydrodynamic pressure they feel at their shock front becomes greater and greater. Our studies of fireballs as they interact with the atmosphere reveal that centimetre-sized rocks undergo progressive fragmentation leading to large fulgurations. Meteoroids of cometary origin, as more fragile aggregates, tend to fracture abruptly when the hydrodynamic pressure exceeds the consistency of these materials, usually in their final stretch. As a consequence, the fine-grained and presumably more volatile interstitial materials are suddenly exposed to the high temperatures and bright flares occur, as I explained above. We can imagine that, given the diameter scale, a debris pile with components tens of metres in diameter would produce an elongated incandescent projectile that would extend its destructive power (Boslough and Crawford, 2008).

There is no doubt that an atmospheric outburst over a populated region could lead to an unprecedented catastrophe (Fig. 4.2). This represents only those extreme cases where the fireball produced in the progressive ablation of an asteroid or comet could reach deep into the Earth's surface and thermally alter its surface. They could be much more dangerous than a direct impact and would cause some tektites. While we should hasten to point out that there is no evidence in our written history of such

Fig. 4.2 Atmospheric outbursts represent only those extreme cases where the fireball reaches deep into the Earth's surface (courtesy of D. Davis)

phenomena, it is also clear that, possibly, such an event may leave no witness to the phenomenon given the very high temperatures that are reached. Certainly, observed from a great distance (hundreds of kilometres away), the fireball could be observed, but there would be no record of the destructive magnitude of the event wherever it occurred. Therefore, they may or may not have gone unnoticed in history.

We might think that talking about these issues borders on science fiction and that it is not possible to prove such events in Earth's history. We would be wrong, because *markers* of the abrupt ramming of large extraterrestrial objects into the Earth have been identified, as we will see in the final chapter. In addition, atmospheric outbursts would produce layers of molten material on the Earth's surface, such as certain tektites found in Thailand (Wasson, 2003). It is possible to find chemical anomalies in such strata because certain chemical elements present in these projectiles are scarce at the Earth's surface. Certain elements, for example, being metal-like, sank with metals during the chemical segregation phase in the early molten phase of our planet. A large collision with an undifferentiated object throws large amounts of ash into the upper atmosphere, which will contain some of the projectile materials. This is why the geological strata containing sediments of the dust released in an impact are often rich in iridium, as I will explain in the final chapter.

4.3 Are These Atmospheric Airbursts Recorded in History?

There are some claims that we have evidence of such an unusual airburst event in the Old Testament, although a real consensus is unclear. Specifically, the Book of Genesis says that the fire rained down on the cities of Sodom and Gomorrah as punishment for their inhabitants' sins. Archaeologist Philip Silvia revealed the results of his excavations in a circular plain north of the Dead Sea, probably where Sodom and Gomorrah were believed to have once stood (Silvia, 2017). In any case, considering that the excavated village called Tall el-Hammam is the biblical Sodom is disputed by biblical scholars and archaeologists. On the other hand, the radiocarbon dating and the origin of the minerals collected in the location taken as a proof of instant crystallization by high temperatures is unfounded for experts in airbursts. The work is controversial for some impact experts, and geologists argue a lack of evidence for such a highly destructive event.

In any case, the debate is served. Alan Bond and Mark Hempsell from Bristol University have studied cuneiform writing on a clay tablet that has puzzled researchers, and that they think is describing an asteroid impact in 3123 BC in Austria. It seems that it might be a contemporary observation of an Aten-type asteroid that impacted Köfels in Austria in the early morning of 29 June 3123 BC. The authors interpret that the back impact plume from the explosion would be bent over the Mediterranean Sea, and re-entered to produce the facts described in Sodom and Gomorrah. Anyhow, this controversial interpretation still needs further research, but it provides a good example of the difficulty in studying such events that occurred in the past, with scarce scientific evidence and using texts of very difficult interpretation. We must be very cautious with those hypotheses until they are fully tested.

In general, when we want to get real evidence of impact hazard, we can refer to the learnings from impact craters dug into the Earth's surface. Large craters have associated distal impact ejecta layers that extend far away from the impact structures and also provide indisputable evidence, like, for example, tektites (Glass and Simonson, 2013). The number of craters is of course biased by the various geological processes that erode their structures and that, over hundreds or billions of years and as a consequence of plate tectonics, virtually erase them completely (Melosh and Ivanov, 1999).

In addition to impact craters, there is another phenomenology that does not excavate craters but can alter the materials on the surface. These are atmospheric outbursts, which occur when the projectile disintegrates in the atmosphere but is so large and massive that the fireball phase in the form of a hot plasma reaches the Earth's surface. Today, this phenomenology is explained by the fact that we know that certain structurally fragile bodies are unable to penetrate until they impact the ground and excavate a crater. Most of them disintegrate long before they hit the ground and generate a column of gas at thousands of degrees in which the object is pulverised. Depending on the geometry, such objects can range from several hundred metres to kilometres in size. As they rapidly deepen in the atmosphere, they experience an increase in hydrodynamic pressure (P) in their frontal region. This pressure is induced by hypersonic entry into the atmospheric layers, which progressively increases their density. It is defined by the average density of the atmosphere and the velocity of the object v according to the equation: $P = \rho \times v^2$. Thus, these objects eventually disintegrate when the consistency of the material is exceeded.

There is clear and fairly recent evidence of such atmospheric outbursts. For example, a carbon-rich black layer about 12,900 years old has been identified at several sites in North America. It appears to coincide with the Younger Dryas, a period marked by the extinction of mammoths and rapid human adaptation to a new and apparently more hostile climatic environment. This period has been explained as an atmospheric outburst by Firestone et al. (2007) on the basis of the discovery of certain extraterrestrial signatures in the sediments, such as the presence of iridium-rich grains, magnetic grains, magnetic microspherules, nanodiamonds and so-called fullerenes. Although this is still under discussion, it may become more precise in the future.

Today we suspect that comets with short orbital periods are the most dangerous objects in this respect. These objects seem to originate from larger, progressively disintegrating objects (Jewitt, 2008). They could therefore be objects that disintegrate progressively and be potential producers of atmospheric outbursts that do not produce craters.

We could ask ourselves some worrying questions: To what extent have these outbursts been recorded in the literature? Wouldn't we have biased information about these events if they had resulted in the local destruction of a civilisation? Could we find evidence of such celestial calamities recorded in ancient documents?

4.4 Asteroids and Comets in the Near-Earth Region

The apparent immutability of the heavens in a human lifetime makes us think that interplanetary space is empty, but we are wrong. Observers of the heavens know that from time to time a meteor or even a fireball may streak across the sky. We just have to wait long enough. This is not surprising since the *Near-Earth region* is traversed by asteroids mostly coming from the main belt and generated over billions of years as a result of multiple collisions (Bottke et al., 2005). Undoubtedly, it also contains a tiny proportion of inactive cometary nuclei, although given their low reflective capacity (albedo) we only know of about a hundred in the near-Earth region. This is why a distinction is made between NEOs (near-Earth objects) and NEAs (near-Earth asteroids) which we refer to intrinsically as asteroids. This is a subtle but important point, given that a small proportion of NEOs are extinct comet nuclei with orbits and reflective characteristics to support this.

Thus, we usually reserve the acronym NEA for true asteroids and NEC for near-Earth comets. Thus, for example, in February 2021 we knew 25,170 NEOs; among them 25,057 are asteroids and 113 follow typical cometary orbits, so they must be comets in their final evolutionary phase. Due to multiple perihelion passes and

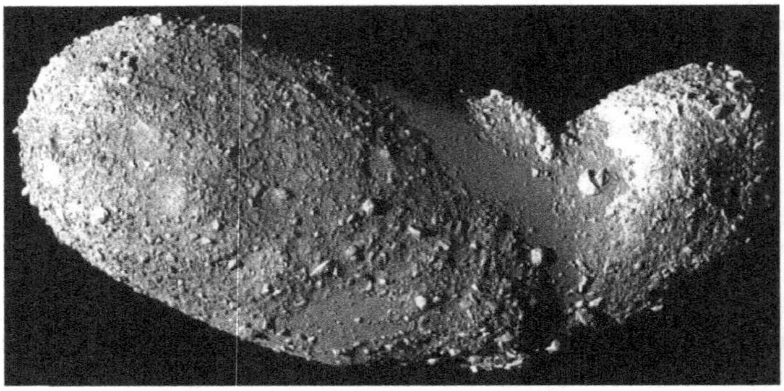

Fig. 4.3 The near-Earth asteroid 25,1433 Itokawa is an obvious example of a debris pile, produced by the catastrophic destruction of two objects and their subsequent gravitational pile-up. Image obtained by the Hayabusa probe (ISAS/JAXA)

close approaches to the planets that subject them to gravitational tidal forces, such short-period comets are doomed to disintegration. Precisely for these reasons, close-in comets (NECs) constitute less than 1% of all known NEOs as revealed by the updated statistics of NASA's *Near-Earth Object Program.*

Space agencies are increasingly interested in studying these objects. It is obvious that if we wish to deal with their hypothetical threat, we must have a good knowledge of the structure and composition of these objects. Therefore, their exploration is not only about their intrinsic scientific teachings or their mining application, but also about their physical properties, in order to be efficient in the event that mitigation methods need to be applied in case they pose a real danger.

Missions aimed at better understanding asteroids and bringing samples back to Earth have already begun. The Japanese space agency Hayabusa landed in November 2005 on NEA 25,143 Itokawa (Fig. 4.3). On 13 June 2010, this probe succeeded in returning samples of micrometre-sized dust particles to the surface of the asteroid in an airtight capsule. These were the first samples ever collected from the surface of an asteroid in the history of space exploration (Nakamura et al., 2012).

4.5 Comet Encke and the Taurid Complex

Fortunately, tracking programmes have made the population of near-Earth objects with a diameter of more than a hundred metres well known today. The region close to our planet is often visited periodically by dark objects of such dimensions and orbital characteristics that relate them to successive fragmentations of a periodic comet. These objects are usually less than 1% of all those known to be approaching the region crossed by our planet and constitute about a hundred near-Earth comets.

The population of extinct comets in the vicinity of the Earth could be an order of magnitude higher for objects with a diameter of less than a hundred metres. The

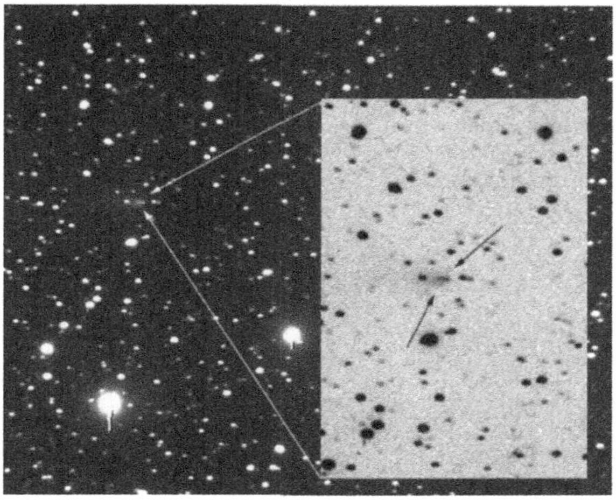

Fig. 4.4 Many periodic comets disintegrate suddenly, as happened to Comet 51P/Harrington. Image of 19 October 2001 from the Begues Observatory in Barcelona, marking the two main nuclei (courtesy of P. Manteca/MPC170)

reason for this would be the catastrophic fragmentation that these objects undergo as a consequence of the gravitational tidal effect of their approaches to the planets or the weakening of their consistency due to the sublimation of ices and organic matter. As comets are aggregates of low-consistency blocks, they fragment into blocks that are small and dark enough to go unnoticed by our tracking programmes (Fig. 4.4). Moreover, their orbits tend to evolve very rapidly, so that on scales of a few thousand years they lose their affinity with the remains or remnant of the comet or asteroid from which they disintegrated.

We can therefore imagine the detective work involved in establishing the existence of multiple objects produced as a consequence of the break-up of one of these short-period comets. The disintegration of these objects, being heterogeneous aggregates, produces smaller objects, often without cometary activity and therefore extremely difficult to detect. Furthermore, to further complicate the resulting product, these disintegrations occur in progressive steps that account for the gradual collapse of the object. However, there is clear evidence of these processes from the study of so-called meteoric complexes. These are the coexistence of multiple asteroids following relatively similar orbits and generally associated with one or even several meteoroid swarms intersecting the Earth's orbit. Among these complexes, one of the best known is the Taurid swarm that is associated with the short-period Comet 2P/Encke and about 20 asteroids as well as several fragment swarms (Babadzhanov et al., 2008) (Table 1). This rock swarm is responsible for several meteor showers intercepted by the Earth today. These showers produce large fireballs that betray the presence of rocks within this swarm (Table 4.1).

Table 4.1 Some of the largest asteroids and comets dynamically associated with the Taurid complex. The diameters given were approximated using various indirect techniques

Object	Semimajor axis (AU)	Diameter (km)
2212 Hephaistos	2.159	6.6
2P/Encke	2.218	4.8
16,960	2.199	4.1
4183 Cuno	1.982	3.8
4341 Poseidon	1.835	2.3
2201 Oljato	2.175	2.2
4486 Mithra	2.200	1.9
2101 Adonis	1.874	0.6

4.6 Was the Tunguska Event a Warning from Comet Encke?

The Tunguska event has been one of the most speculated-about events in the literature. It took place on 30 June 1908 and, despite occurring over a practically uninhabited area of Siberia, aroused unusual media attention given the detection of seismographs and the effects it caused on the Earth's atmosphere. Among the far-fetched hypotheses that have been proposed to explain it, some as surprising as a collision with a mini-black hole, antimatter particles, an atomic bomb or even an extraterrestrial spacecraft could be listed. None of these hypotheses took into account that the description of the phenomenon and its immediate effects—gas plume visible at great distances, seismic waves induced by the shock wave, massive amounts of dust in the stratosphere, etc.—clearly support the idea of a meteoric event.

It wasn't until 1930 that the English meteorologist Francis J.W. Whipple (1876–1943) suggested that the object which produced the Tunguska event must have been a small comet. Other researchers such as David Hughes and the American astronomer Fred Whipple have supported this hypothesis (Whipple, 1975). The hypothesis was gaining force until a more detailed study by Kresak (1978) suggested that the Tunguska event could have been produced by a fragment of Comet 2P/Encke. After all, the hypothesis could be a natural consequence, as we know that the evolution of short-period comets ends in its disintegration (Hughes, 2002). In addition, regarding that encounter on 30 June, it coincides with the dates when the Earth crosses a meteoroid swarm associated with that comet: the Beta Taurids. The estimated geometry of the fireball's trajectory shows that it could have made a significant deviation (about 20°), but it has been suggested that this effect could be a consequence of non-gravitational perturbations on its heliocentric orbit (Asher and Steel, 1998)—this is the name given to the sum of physical processes that, in objects of a few hundred metres, most effectively induce the progressive loss of orbital energy and the consequent decrease in the orbital period of the object.

At present, there is still no clear consensus on the origin of the Tunguska event, but the association with Comet 2P/Encke seems feasible (Kresak, 1978; Asher and Steel, 1998). The idea that it was a low-consistency object, i.e. cometary in nature,

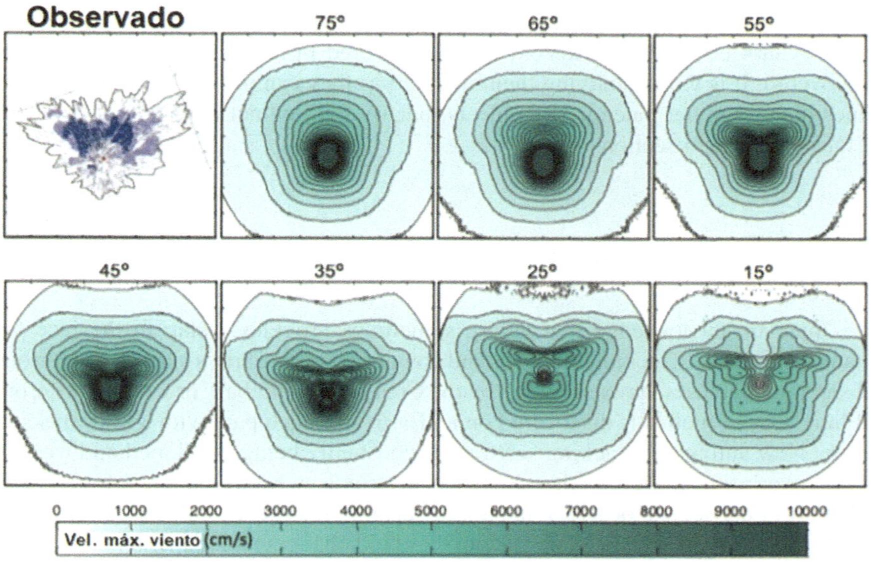

Fig. 4.5 The pattern of fallen trees has been used to calculate the trajectory and angle of entry of the Tunguska event (courtesy of M. Boslough)

has recently been reaffirmed by physical models indicating that it may have had an initial mass of 8 million tonnes, capable of going down to about 10 km above the Earth's surface and depositing a total energy of 10 megatonnes of TNT (Gladysheva, 2020a, b).

The possible trajectory could also be improved by employing an ingenious method: using the pattern of fallen trees that depend on the disintegration site of the object and the propagation of the shock wave as a function of the maximum wind speed (Fig. 4.5). Thus, it has been determined that the entry angle (around 32°) of the Tunguska superbolide would support the hypothesis that it was associated with the Beta Taurid swarm produced by Comet 2P/Encke (Boslough and Crawford, 2008; Boslough and Brown, 2018). Obviously this does not mean that the association is entirely clear, since it is not possible to know the initial velocity of the object in sufficient detail to be able to model its heliocentric orbit with any certainty. This should not be surprising, as it is also difficult in the case of recording the fireball from multiple casual videos and satellite data in the case of the Chelyabinsk asteroid (Borovička, et al., 2013) or the Villalbeto de la Peña meteorite.

Whatever it was, a Tunguska-like event could happen again at any time. Space missions have revealed that the difference between asteroids and comets seems much more subtle than was thought a few decades ago. One example is provided by the different asteroids that make up the Taurids complex, indicating that other potential Tunguska (asteroids several tens of metres in size) may exist in similar orbits and have not been discovered to date. Indeed, it has been proposed that these bodies might be best discovered under certain conditions, given that they follow resonant orbits (Clark et al., 2019). Recently, comets stored in the asteroid belt have been discovered. It is worth studying them in detail as they can provide us with valuable clues about another category of objects that we call extinct comets, which I will discuss in the next section.

4.7 A Great Pumpkin for Halloween: 2015 TB145

It is hard to acknowledge that nature sometimes provides a slap in the face for mankind's hubris. Tracking asteroids and comets from the Earth's surface has serious observational biases, as exemplified by the Chelyabinsk fireball, or when an asteroid of hundreds of metres is discovered on an Earth approach path with a few weeks to spare. New automated tracking software is constantly being developed and we will soon see orbiting telescopes that can detect asteroids by exploiting the infrared band where they are several orders of magnitude brighter. However, despite our efforts to catalogue these bodies, a chance encounter could change human history.

Sometimes long-period comets cross paths with the almighty Jupiter, the most massive and influential star in the solar system, except for the Sun. Jupiter can be considered our planet's great protective ally (Trigo-Rodriguez, 2012). The outcome of a comet's boldness in crossing Jupiter's orbit varies depending on the proximity of the encounter. If the comet gets too close, it can be trapped by Jupiter's gravitational field and then gradually fall towards it. Such a case is exemplified by Comet Shoemaker–Levy 9. It was discovered in the vicinity of the planetary giant on 25 March 1993 by Eugene and Carolyn Shoemaker with David Levy, in photographs taken with the 46 cm Schmidt telescope at the Palomar Observatory (California). Its appearance was unusual in that it was a multiple comet, consisting of a dozen or so aligned nuclei. It was a comet in the midst of fragmentation that later became at least 21 fragments. But that had not always been the case. Orbital analysis revealed that the comet, with an estimated initial diameter of 1.6 km, had undergone a close encounter with Jupiter around 1929 that had captured it in a jovicentric orbit. It remained in this orbit for several decades until 8 July 1992, when it passed within 22,100 km of the planet's high clouds and fragmented under the effect of gravitational tidal forces. As a consequence, the fragments were distributed in highly eccentric orbits that, when gravitationally perturbed by the Sun, would eventually reduce their minimum orbital distance to Jupiter until the impacts occurred between 16 and 18 July 1994 (Chapman, 1993; Noll et al., 1995).

This encounter was a scientific milestone of the first magnitude. The scientific community indirectly corroborated what was already suspected: comets can be very fragile aggregates, even piles of debris (Solem, 1994). This led to experiments in terrestrial laboratories to simulate the unique materials and properties seen in comets (Blum et al., 2006). As a result, the idea was reaffirmed that these objects from the outer regions of the solar system may be extremely primordial, just-compacted aggregates of the materials that were consolidated in the outer regions of the protoplanetary disc more than 4.565 billion years ago (Brownlee et al., 2006).

In many cases, however, the object is not retained, but fragments into multiple pieces whose orbits are shortened. Such objects often have considerable eccentricity and move away from the Sun to just under the orbit of Jupiter, precisely because it is often involved in their fragmentation by the gravitational tidal effect it confers on objects of this type that come too close to it. Such objects may even have ceased to behave like comets, since in fragmentation they extinguish much of their volatile materials and, moreover, in their new orbits they will pass more often in the vicinity of the Sun, where they would extinguish their volatiles or cover their surface with regolith. In addition these objects are subjected to space weathering processes that change the reflectance properties of these bodies (Chapman, 2004; Tanbakouei et al., 2020)

These hypotheses are being supported by observational evidence. On 10 October 2015 a faint object called 2015 TB145 was discovered (see Fig. 4.6). That day we became collectively aware of the existence of extinct comet nuclei with worrying peculiarities: dark, extremely fast due to their high orbital eccentricity, and unexpected. That extinct comet had a very low surface reflectivity, what astronomers call *albedo*, barely 5% in the visible range. A direct consequence of its faintness, and despite its size, the object was discovered with only three weeks to spare. Fortunately, that monster passed our planet at a distance slightly greater than that which separates us from our satellite: 1.27 lunar distances. We don't know exactly how many of these eccentric bodies are out there, but we should be spurred to create dedicated space telescopes capable of discovering them all. A telescope using infrared detectors could

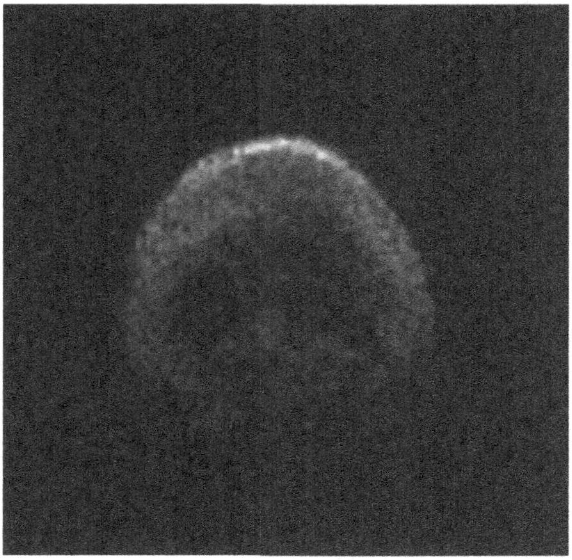

Fig. 4.6 The strange appearance of asteroid 2015 TB145 obtained on 2 October 2018 from the Arecibo radio telescope in Puerto Rico (NAIC-Arecibo/NSF)

be the best way, as these objects could be in that photometric band several orders of magnitude brighter than in the visible range.

This potentially dangerous object, classified as an Apollo group object because of its orbital characteristics, suggests that perhaps we underestimate the extreme danger associated with this type of large, dark and fast-moving extinct comet fragment. These objects exemplify, in my opinion, the 'dark side' of the dangers of impact with the Earth. Due to the eccentric orbit of 2015 TB145 it has a speed of about 35 km/s (126,000 km/h) and would therefore hit our planet with about ten times the energy of an asteroid arriving from the main belt. And, at around 650 m in diameter, we are talking about an object whose mass would exceed 300 million tonnes. A monster truly capable of producing regional or even global devastation if it were to impact at such speed. Such an event would be catastrophic on a global scale, possibly producing an era change like that caused by the asteroid which excavated the Chicxulub crater in the Gulf of Mexico some 65 million years ago.

Questions for Chapter 4

Q.1. Astronomers are puzzled by big questions that last decades or centuries. If comets were formed at long distances from fragile aggregates that produce challenging properties like, for example, small bulk density and tensile strength, why can some of them reaching the inner solar system disrupt many rocky bodies like e.g. 2P/Encke and the Taurid complex? Try to answer by giving your own arguments. [Solution: The answer to this question probably still needs to be written, and should be studied case by case after a careful study of the bodies. In any case, some of these short-period comets could have experienced particular histories: encounters with other small bodies, reaggregation after collisions with asteroids adopting a rubble-pile like structure, planet gravity-assisted disruptions and scattering, etc.…].

Q.2. It is still matter of debate, but some researchers think that the Tunguska event was produced by a 50-m fragment of a comet, perhaps a piece of the Taurid complex. Comets are fragile bodies composed of volatile species that might cause an extremely large fireball. If the ablation column touches the ground the effects are devastating. About 2,150 km^2 of Siberian taiga was destroyed by the shock wave, and a significant part was incinerated as a consequence of the fires produced by the hot gas reaching the ground. Try to calculate the probability of any particular city on Earth being destroyed by such a disruptive event. To simplify, you can assume that the entire Earth surface can be obtained from the surface area of a sphere ($4\pi \cdot r^2$). [Solution: Dividing the total surface of our planet by the total known devastated area in Tunguska we can get a first order of magnitude approach to the probability of a specific area of that size being hit for a type of event. Then, 5.1×10^8 km^2 / 2150 km^2 gives 237,255, so apparently the probability is 1 in 237,255. If we consider that only about two-fifths of the Earth's surface is covered by land, then we can recalculate the probability as: $237{,}255 \times 2/5 = 94{,}900$. About 1 in 100,000. If we extend the calculations for a big city a few tens of km across the probability of being hit is about one order of magnitude minor.]

Q3. Most short-period comets end their life as 'dormant' comets that become dark, almost inactive, asteroids. The reason for such behaviour is because the heaviest particles cannot be ejected by the gas drag sublimated so they fall into the surface where they accumulate. Consequently, they end covered by rubble mantles, refractory materials that are left behind the cometary surface due to the extended cometary activity during hundreds of perihelion approaches. Try to estimate the maximum size of a meteoroid released from a spherical comet 10 km in diameter, located at 1 AU from the Sun, exhibiting a sublimation rate per unit of area: dm/dt = $10^{-4\ \mathrm{kg\ m}-2\ \mathrm{s}-1}$ and experiencing an outgassing gas velocity (V_s) as a function of the heliocentric distance (d) that can be written as follows in first approach: $V_s = 500d^{-1/2}[inAU]$.

[Solution: The surface gravitational force (g) for a meteoroid of radius r, can be written as a function of the gravitational constant G (6.6×10^{-11} $\mathrm{Nkg^{-2}m^2}$), ρ and R are the cometary nucleus bulk density, and its radius, respectively:

$$g = \frac{16}{9}\pi^2 G \cdot \rho \cdot R \cdot r^3$$

The gas drag force is a quite complex term as it depends on the shape of the particle, and of the ratio between the particle size and the mean free path in the evolving sublimated gas. We can simplify the problem assuming that the size of the meteoroid is much smaller than the mean free path in the gas. We can reach the drag force equation as follows (for further details see Jewitt, 2008):

$$F = C \cdot \pi \cdot r^2 \cdot \mu \cdot m_H \cdot N \cdot \Delta V^2$$

That equation contains the drag coefficient (C), the molecular weight of the sublimating gas (μ), the mass of the hydrogen atom (m_H), the concentration of the gas at the level of the surface (N), and the velocity of the gas relative to the grain (ΔV).

Then, by balancing the gravitational force experienced on a spherical meteoroid with a density (ρ_m) with the gas drag we will reach an equation for the critical size for which it must fall into the cometary surface:

$$r_{critical} = \frac{\mu m_H N V_s}{G\rho\rho_m R}$$

The previous equation for the critical radius of the meteoroid can be rewritten (as dm/dt $= \mu m_H N V_s$) as follows:

$$r_c = \left[\frac{V}{G\rho\rho_m R}\right]\frac{dm}{dt}$$

Considering the initial values given in the problem, we will find that the maximum size for a meteoroid released in those circumstances is about 10 cm. Given the approaches made in the calculations, the result seems to be consistent with most observations of large bolides reaching the top of the atmosphere that are a few tens of cm in diameter (Trigo-Rodríguez and Blum, 2021). Those reaching the Earth from fast-encounter velocity meteoroid streams, like the Perseids or Leonids, have enough kinetic energy to produce bolides as bright as the full Moon.]

Online Resources

Asteroid Day: https://asteroidday.org/

Asteroid Damage Visualization Map: https://asteroidcollision.herokuapp.com/

Earth Impact Effects Program:
https://impact.ese.ic.ac.uk/ImpactEarth/ImpactEffects/

References

Asher D. and D.O. Steel (1998) On the possible relation between the Tunguska bolide and Comet Encke. *Planetary and Space Science* 46, 205–211.

Babadzhanov P.B. et al. (2008) Near-Earth Objects in the Taurid complex. *Monthly Notices of the Royal Astronomical Society* 386, 1436–1442.

Bland P. and N.A. Artemieva (2003) Efficient disruption of small asteroids by Earth's atmosphere. *Nature* 424, 288–291.

Blum J., R. Schräpler, B.J.R. Davidson and J.M. Trigo-Rodríguez (2006) "The physics of protoplanetesimal dust agglomerates. I. Mechanical properties and relations to primitive bodies in the solar system", *Astrophysical Journal* 652, 1768-1781.

Borovička, J., P. Spurný, P. Brown, P. Wiegert, P. Kalenda, D. Clark and L. Shrbený (2013) The trajectory, structure and origin of the Chelyabinsk asteroidal impactor. *Nature* 503, 235.

Bottke Jr. W.F. et al. (2005) Linking the collisional history of the main asteroid belt to its dynamical excitation and depletion. *Icarus* 179, 63–94.

Boslough M. and D.A. Crawford (2008) Low-altitude airbursts and the impact threat. *International Journal of Impact Engineering* 35, 1441–1448.

Boslough M. and P. Brown (2018) Was Tunguska a Beta Taurid? *2019 Observational Campaigns can Test Hypothesis. American Geophysical Union, Fall Meeting 2018*, abstract #P53D-2998.

Brownlee D. et al. (2006) Comet 81P/Wild 2 under a microscope. *Science* 314, 1711–1716.

Cabedo V., J. Llorca, J.M. Trigo-Rodríguez, A. Rimola (2021) Study of Fischer-Tropsch Type reactions on chondritic meteorites, *Astronomy & Astrophysics* 650, A160, 23 pp.

Chapman C.R. (1993) Comet on target for Jupiter. *Nature* 363, 492–493.

Chapman C. (2004) Space weathering of asteroid surfaces. *Annual Review of Earth and Planetary Sciences* 32, 539–567.

Clark D.L., P. Wiegert and P. G. Brown (2019) The 2019 Taurid resonant swarm: Prospects for ground detection of small NEOs. *Monthly Notices of the Royal Astronomical Society* 487, L35–L39.

Firestone R.B. et al. (2007) Evidence for an extraterrestrial impact 12,900 years ago that contributed to the megafaunal extinctions and the Younger Dryas cooling. *Proceedings of the National Academy of Science* 104, 16016–16021.

Gladysheva, O.G. (2020a) Swarm of fragments from the Tunguska event. *Monthly Notices of the Royal Astronomical Society* 496, 1144–1148.

Gladysheva, O.G. (2020b) The Tunguska event. *Icarus* 348, 113837.

Glass B.P. and B.M. Simonson (2013) *Distal Impact Ejecta Layers*. Springer-Verlag, Berlin Heidelberg, Germany, 716 pp.

Hsieh H. and D. Jewitt (2006) A population of comets in the Main Asteroid Belt. *Science* 312, 561–563.

Hughes D.W. (2002) The magnitude distribution and evolution of short-period comets. *Monthly Notices of the Royal Astronomical Society* 336, 363–372.

Jewitt D. (2008) Kuiper Belt and Comets: An observational perspective. In *Trans-Neptunian Objects and Comets*, Swiss Society for Astrophysics and Astronomy, Springer-Verlag, Heidelberg, Germany, pp. 1–78.

Kresak L. (1978) The Tunguska object: A fragment of Comet Encke? *Bulletin of the Astronomical Institutes of Czechoslovakia* 29, 129–134.

Melosh H.J. and B.A. Ivanov (1999) Impact crater collapse. *Annual Review of Earth and Planetary Sciences* 27, 385–415.

Nakamura E. et al. (2012) Space environment of an asteroid preserved on micrograins returned by the Hayabusa spacecraft. *Proceedings of the National Academy of Sciences.* pnas.1116236109

Napier B. and D. Asher (2009) The Tunguska impact event and beyond. *Astronomy & Geophysics* 50(1), 18–26.

Noll, K.S., M.A. McGrath, L.M. Trafton, S.K. Atreya, J.J. Caldwell, H.A. Weaver, R.V. Yelle, C. Barnet C. and S. Edgington (1995) HST spectroscopic observations of Jupiter following the impact of Comet Shoemaker–Levy 9. *Science* 267, 1307–1313.

Rotelli L., J.M. Trigo-Rodríguez, C.E. Moyano-Cambero, E. Carota, L. Botta, E. Di Mauro and R. Saladino (2016) The key role of meteorites in the formation of relevant prebiotic molecules in a formamide/water environment, *Nature Scientific Reports* 6:38888, doi: https://doi.org/10.1038/srep38888.

Silvia P.J. (2017) *The civilization-ending 3.7KYrBP Event: Archaeological data, sample analysis, and biblical implications.* 80th Annual Meeting of the Meteoritical Society, held July 23–28, 2017 in Santa Fe, New Mexico. LPI Contribution No., 1987, id.6001

Solem, J.C. (1994) Density and size of Comet Shoemaker–Levy 9 deduced from a tidal breakup model. *Nature* 370, 349–351.

Tanbakouei S., J.M. Trigo-Rodríguez, J. Blum, I. Williams and J. Llorca (2020) Comparing the reflectivity of ungrouped carbonaceous chondrites with that of short period comets like 2P/Encke, *Astronomy & Astrophysics* 641, A58, 7 pp.

Trigo-Rodríguez J.M. (2012) *Las raíces cósmicas de la vida, Colección El Espejo y la Lámpara.* Servei Publicacions UAB, Barcelona.

Trigo-Rodríguez J.M. and J. Blum (2021) Learning about comets from the study of mass distributions and fluxes of meteoroid streams. *Monthly Notices of the Royal Astronomical Society*, stab 2827, 13 pp.

Trigo-Rodríguez J.M. et al. (2014) UV to far-IR reflectance spectra of carbonaceous chondrites. I. Implications for remote characterization of dark primitive asteroids targeted by sample-return missions. *Monthly Notices of the Royal Astronomical Society* 437, 227–240.

Trigo-Rodríguez J.M., A. Rimola, S. Tanbakouei, V. Cabedo-Soto, and M.R Lee (2019) Accretion of water in carbonaceous chondrites: current evidence and implications for the delivery of water to early Earth, *Space Science Reviews* 215:18, 27 pp.

Trigo-Rodríguez J.M., J. Dergham, M. Gritsevich, E. Lyytinen, E.A. Silber and I.P. Williams (2021) A numerical approach to study ablation of large bolides: Application to Chelyabinsk. *Advances in Astronomy* 2021, 8852772.

Wasson J.T. (2003) Large aerial bursts: An important class of terrestrial accretionary events. *Astrobiology* 3, 163–179.

Whipple F. (1975) Do comets play a role in galactic chemistry and y-ray bursts? *Astronomical Journal* 80, 525–531.

Chapter 5
Quantifying the Nature and the Magnitude of the Hazard from Asteroid Impacts

One of the aspects of space exploration that may be of most concern is the growing evidence that the near-Earth region is being traversed by capricious visitors from outer regions of the solar system. Considerable effort has recently been devoted to understanding this population of near-Earth asteroids. The challenge is of a magnitude that can be exemplified by the following example. We know that asteroids pass close to our planet, but even for those that come closest we were until recently unable to quantify their true dimensions. The problem is that, even if they pass relatively close on an astronomical scale, their dimensions cannot be resolved with our largest optical telescopes.

Until recently, we were not really aware of the real magnitude of the hazard associated with asteroids and comets. In fact, it wasn't until the return of 1P/Halley's Comet in 1910 that humanity started to think about these approaches to our planet. At that time, our knowledge about comets was rich enough to dismiss any risk, but some parts of the media emphasized that our planet was crossing the comet tail. Many unofficial press reports emphasized the notion that this could have consequences, due to the poisonous gases within. Obviously, the tenuous nature of a comet ion tail cannot enter our dense and magnetically protected atmosphere, but the speculations alarmed a part of the population. While astronomers were trying to calm the people, others were making a fortune selling 'anti-comet pills'. In any case, the popular perception of Comet Halley's approach to Earth was also magnified by other circumstances. For example, the casual appearance of the Great Daylight Comet of 1910, which surpassed Halley in brilliance, being brighter than planet Venus. This long-period comet exhibited a long dust tail that made it really noticeable in broad daylight, just about four months before Comet Halley made its appearance.

Concerning the risk associated with asteroids, a major breakthrough was the discovery of the first Apollo asteroids, starting with 1862 Apollo, discovered in 1932 by the German astronomer Karl Reinmuth (1892–1979). The orbital characteristics of the Apollo-type NEAs suggested to many astronomers the relevance of identifying them as a source of risk. The first modern asteroid that was recognized as a source of risk to humans was asteroid 69,230 Hermes, with its close approach

J. M. Trigo-Rodríguez, *Asteroid Impact Risk*, Impact Studies,
https://doi.org/10.1007/978-3-030-95124-5_5

to 741,000 km from the Earth on 30 October 1937. In mythology, Hermes was the messenger of the gods, son of Zeus and Maia, so its name was somehow appropriated to gain awareness about impact hazard.

I will discuss below the ingenious techniques that allow us to reconstruct the shapes and know their true dimensions, before emphasizing that, from telescopic observations and the apparent magnitude of these objects, we can only approximate these values with a non-negligible error rate.

5.1 Shaping the Sources of Hazard

Pioneer studies of the reflectance spectra of asteroids compared with the ones obtained from meteorites make possible to identify the main rock-forming materials of these bodies (see e.g. Pieters and McFadden, 1994). In addition, telescopic images can give a sense of the dimensions of these objects based on their magnitude of brightness related to an equivalent diameter (Bottke et al., 2000, 2002a). The magnitude at which we see each asteroid in the field of our telescope depends on many factors, but are mainly: its shape, phase angle, chemical composition, relative angular velocity and its reflective capacity (given by the *albedo* or ratio of light reflected/received by the asteroid). Although it is possible to obtain spectra of these asteroids and compare them with those of meteorites—and even to reconstruct their approximate shapes based on the study of their light curves—it is apparent that the task is enormous. Unlike the main-belt asteroids that contain some large asteroids (Table 5.1), the asteroids in the near-Earth region are much smaller: only about one thousand are larger than 1 km across. In fact, the smaller ones, ranging from metre- to kilometre-sized, due to their smaller mass are more affected by the so-called non-gravitational effects that influence their progressive drift and cause them to cross orbital resonance zones that can transfer them from one orbit in the main belt to a much more eccentric orbit in the region of the terrestrial planets.

Today, the enormous sensitivity of modern CCD cameras makes it possible to detect extremely fast and fleeting asteroids. The closer they are to the Earth, the

Table 5.1 Asteroids with a diameter greater than 300 km, indicating their diameter, mass and mean density (where known)

Asteroid	Diameter (km)	Mass (kg)	Average density (g/cm^3)
1 Ceres	913	9.5×10^{20}	2.1
2 Pallas	523	2.4×10^{20}	2.9
4 Vesta	520	2.7×10^{20}	3.5
10 Hygiea	429	1.1×10^{20}	–
511 Davida	337	1.1×10^{20}	–
704 Interamnia	333	7.0×10^{19}	–
52 Europe	312	5.2×10^{19}	–

higher their angular velocity in the sky and therefore the more elusive they are. Some of them already trace trajectories in the CCD images of several arc minutes within a few seconds. Obviously, these asteroids went unnoticed decades ago, until the advent of CCD digital-imaging techniques. In fact, these cameras have been a revolution in astronomy. Not surprisingly, historically it was the application of photography in the discovery of hundreds of asteroids. One pioneer was the astronomer Josep Comas i Solà, who began discovering asteroids in 1915 and remained one of the most outstanding discoverers until the 1970s. This pioneer of astronomy, often forgotten because of his free and libertarian nature, was also one of the greatest popularisers of astronomy in his time. He used a technique that he himself developed to demonstrate the proper motion of these stars in comparison to the drift of the stars due to the effect of the Earth's rotation. This was an enormous challenge at the time, as there were no film cameras as we know them, but instead large cameras on large plates. An example is given in Fig. 5.1, which shows the displacement of the asteroid (804) Hispania when it was discovered on 20 March 1915.

Obviously in these images, whether photographic or taken by digital or CCD cameras, the asteroids are mere points of light. They would be indistinguishable from stars, were it not for their own motion around the Sun. A giant step forward involves the use of complementary techniques, even if they are not always applicable, but only in particular cases. In this sense, the closest approaches of asteroids to the Earth allow us to study them using radar techniques, particularly with the largest antennas we have for this purpose: mainly Arecibo and Goldstone.

The enormous size of such radio telescopes is not accidental since, due to their use of longer radio wavelengths, they require high spatial resolutions and thus diameters of several tens or even hundreds of metres. The larger they are, the higher the special resolution. Taking advantage of the hollowness of the valley, the huge Arecibo radio

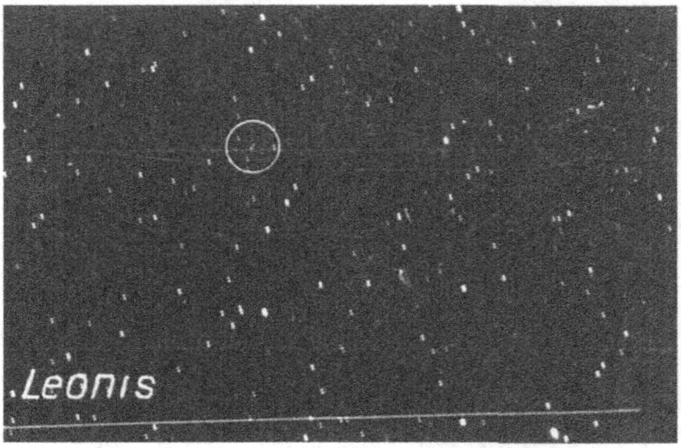

Fig. 5.1 Image of the asteroid (804) Hispania on the original plate of its discovery by Josep Comas i Solà from Villa Urania. He used a stereoscopic method that he himself developed. The image consists of a pair of successive exposures on the same plate. The circle shows the asteroid in the circle which, due to its displacement, has a different motion from the stars (courtesy of Observatori Fabra)

telescope in Puerto Rico with its 305-m diameter had done an impressive job until collapsing in 2020. While we wait for it to be rebuilt, a similar one, called FAST, has been created in China, with a diameter of 500 m. Russia's Academy of Sciences also has a 576-m diameter instrument made up of rectangular radius reflectors that can be found in the Greater Caucasus. The job of these metal monsters in this field will be to continue to provide us with highly detailed images of the irregular structure of the objects that come closest to Earth.

Other techniques, such as asteroid occultations of stars, have also been essential for inferring the shape of such objects. By knowing the exact orbit of an asteroid around the Sun, we can predict when the asteroid will cross a background star. The accuracy of that prediction will depend on how precisely we know the asteroid's heliocentric orbit. Modern video or CCD cameras allow such transits in front of stars to be closely tracked and, if multiple observers observe the occultation from distant locations, the shape of the asteroid at that instant can be reconstructed (Fig. 5.2).

Other studies of the photometric behaviour—i.e. of the asteroid's luminosity as a function of time—allow the approximate shapes of asteroids to be reconstructed on the basis of geometric models. A convolution technique is applied to the luminous flux that is of a variable nature, among other factors due to rotation, the phase angle of

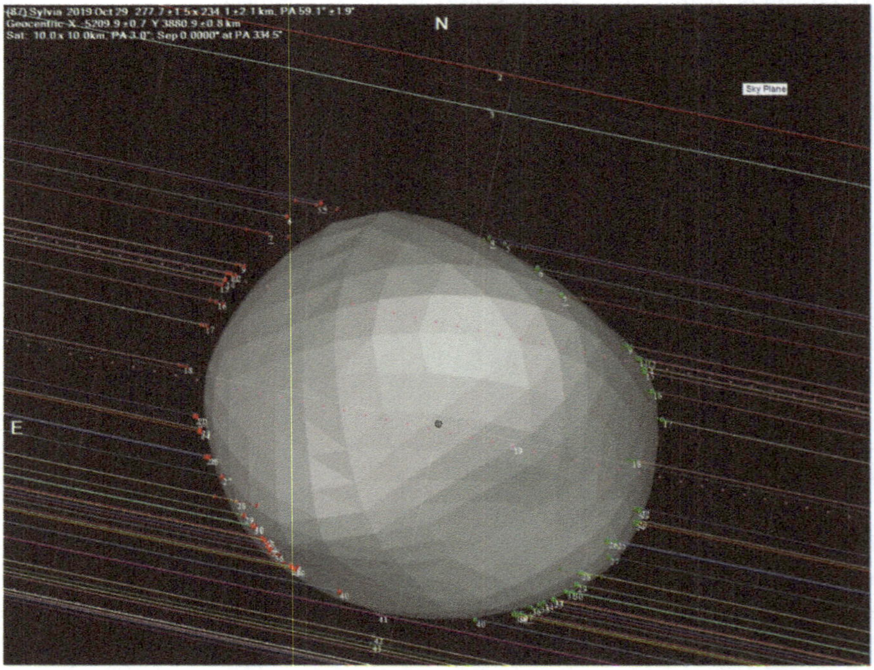

Fig. 5.2 Reconstruction of the silhouette of the asteroid Sylvia using the method of arcs of various stations looking at the occultation of a star behind it. This event occurred on 29 October 2019 and involved more than 50 observers (courtesy of Eric Frappa / EURASTER)

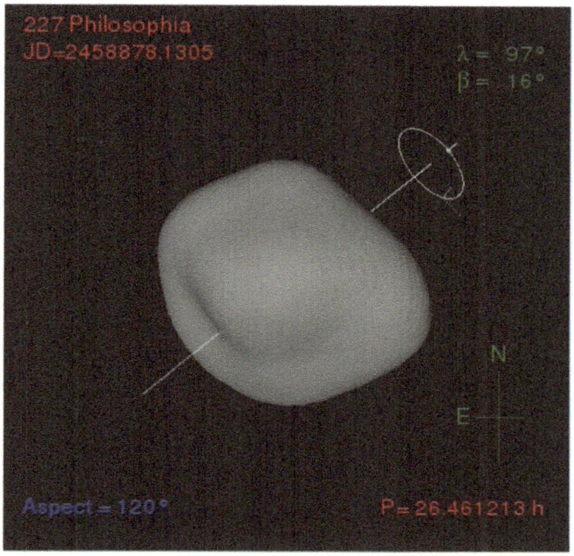

Fig. 5.3 The shape of asteroid (227) Philosophia, reconstructed using the convolution technique (Marciniak et al., 2018)

the object in relation to the Sun, the shape and composition of the asteroid itself. This is a mathematical operator that transforms the light profile obtained as a function of time to reconstruct the shape of the asteroid. The technique has been improved over the last decades and the current models are quite accurate (see example in Fig. 5.3).

Sometimes these objects cross the Earth's orbit, just shy of lunar distance (LD), as if to hint that they could complete their cosmic carom at any moment. In fact, it is only a matter of time, as more and more small asteroids are known to hit our planet with increasing frequency. This awareness begs the questions: To what extent do these objects pose a danger? To what extent have they influenced the evolution of life on our precious planet? Let's take it one step at a time.

A key aspect that needs to be quantified is the local or regional devastation that an asteroid would cause as a function of its diameter. A pioneering work in this regard was that of Hills and Goda (1993), who indicated that objects less than 150 m in diameter and with a density of about 3 g/cm^3 would deposit most of their kinetic energy in the fireball phase in the atmosphere, rather than excavating a crater. This is due to the efficient deceleration and fragmentation of these projectiles, which disintegrate in the fireball phase and generate so-called *airbursts* like the one in Tunguska. It also helps that many of the projectiles reaching Earth arrive heavily shocked and fractured, i.e. skewed by the processes that send them into interplanetary space and subsequently en route to Earth (Popova et al., 2011). A later study, based on the relative absence of small craters on the Earth's surface, placed this upper limit at about 220 m in diameter (Bland and Artemieva, 2003). However, given the complexity of the calculations and the intrinsic diversity of objects that can reach the top of the atmosphere—monolithic, fractured or *rubble piles*—a limit of around 150 m is still maintained.

Indeed, there are several graphs that quantify the level of regional devastation depending on the physical properties of the projectile (Hills and Goda, 1993). For a projectile with a velocity of about 20 km/s, these authors showed that the Earth's atmosphere is no longer effective in fragmenting an asteroid with a diameter exceeding about 100 m, whereas due to the fragility of comets it could reach a diameter of up to 500 m. Obviously metallic meteorites, which are much denser, could excavate a crater even with much smaller diameters of between 20 and 30 m. It would evidently also depend on the relative velocity of the projectile's encounter with the Earth, as well as the angle of incidence, among other factors. Interestingly, although some bodies can be efficiently fragmented by friction and hydrodynamic ablation, the shock wave penetrating at hypervelocity and the burst of hot gas generated (called an *airburst*) could also be very dangerous. If the incandescent plasma cloud we call a fireball hits the ground, it could incinerate all forms of life over a very large region and even melt the surrounding rocks. This is how, for example, some tektites found in the Libyan (Sahara) desert are thought to have been produced.

At this point, the reader will probably realise that prevention of these natural phenomena is the first step to be taken. The last decade has seen a technological revolution that has automated the detection of these dangerous projectiles. Comprehensive cataloguing of the largest asteroids has made it possible to predict their approaches and to study them on the basis of these unusual but momentous encounters.

5.2 Monitoring Programmes

At the end of the twentieth century, the US agency NASA set out to promote a fundamental technological challenge when it came to foreseeing the encounters of dangerous asteroids and comets with the Earth. Knowing that the first step was to obtain the most complete and accurate cataloguing possible of all potential projectiles, several systematic programmes were designed and implemented to detect and orbitally track the vast majority of these near-Earth objects (known as NEOs, from the acronym for Near-Earth Objects) with a diameter of more than one kilometre. These NEOs include asteroids (known as NEAs) and extinct comets (NECs), although the latter make up slightly less than 1% of the total population. A decade later, the milestone of having catalogued 99.9% of objects with a diameter of around one kilometre has been almost completed thanks to various monitoring programmes (see Table 5.2). Nevertheless, we are aware that the programme has significant gaps, since to date the detection has been carried out exclusively from telescopes located on the Earth's surface whose records are biased in certain directions in space due to particular geometries and seasonal parameters, such as the intrinsic observability of certain regions of the celestial sphere due to day–night variations.

As a pioneering follow-up programme, the *Spacewatch* project, established in 1984 under the scientific leadership of Tom Gehrels (1925–2011), has been using the 90-cm-diameter telescope at the Steward Observatory. This programme pioneered the use of electronic detectors (CCD cameras) for asteroid detection. In fact, Gehrels'

Table 5.2 Major monitoring programmes, year of commencement of monitoring operations and the average number of discoveries of NEAs of different diameters (Ø) since the establishment of the programme. Data are counted globally since 1995

Programme name	Starting year	All NEAs	NEAs (Ø > 140 m)	NEAs (Ø > 1 km)	Discovery rate (NEAs/year)
Pan-STARRS	2010	7738	2522	29	645
Catalina	1998/	12,644	3286	154	527
LINEAR	1996/2017	2633	1761	324	120
ATLAS	2015	682	186	2	97
NEAT	1996/2007	444	388	63	37
NEOWISE	2010	341	285	15	28
LONEOS	1998/2008	290	232	40	26
Spacewatch	1984	857	504	48	23

CV includes operating the first system to discover an asteroid and a comet with CCD, and even the first to do so using an automatic image-processing system.

Among the most successful monitoring programmes to date is *LINEAR (Lincoln Near-Earth Asteroid Research)*, an initiative of the Massachusetts Institute of Technology's Lincoln Laboratory, which was funded by the US Air Force and NASA. Since the beginning of its firmament monitoring activity at the end of 1997, it has been characterised by its innovative nature in the application of ultra-sensitive digital technology and, in fact, it eventually became one of the key programmes in meeting NASA's challenge of discovering the majority (90%) of near-Earth asteroids with a diameter of one kilometre or more (Table 5.1).

The *LONEOS* (Lowell Observatory Near-Earth-Object Search) programme began in 1993 using the 60-cm-aperture Schmidt telescope in Flagstaff, Arizona. Using a 4 × 4 CCD detector, it can cover a field of 2.9 square degrees (an angular diameter of ~10 times our Moon). A team from NASA and the Jet Propulsion Laboratory (JPL) reached an agreement in 1996 to use the 1-m GEODSS telescope on Mauna Kea, owned by the US Air Force, for a monitoring programme known as NEAT. Subsequently, in 2000, this programme switched to the AMOS 1.2-m telescope at a nearby site.

Catalina Sky Surveys (CSS) is the name given to three joint programmes for the discovery of NEAs located in Arizona (Steward Observatory and Mount Lemmon) and one in Australia (Siding Spring). It is currently one of the most efficient programmes, having contributed dozens of discoveries per year.

The most popular programme at the moment, possibly because of its innovative use of colour filters for the discovery of NEAs, is known as Pan-STARRS (Panoramic Survey Telescope and Rapid Response System). It began full operation in November 2012 and has already claimed discoveries of famous asteroids and comets. It uses a 1.8-m-diameter telescope located in Maui, Hawaii, whose CCD camera was once the world's largest with a 64 × 64 detector array producing 1.4 gigapixel images.

NEOWISE is a NASA-funded programme to find NEAs in images taken by the Wide-field Infrared Survey Explorer (WISE) probe. It is remarkable that the discovery of asteroids from Spain has been led by the Astronomical Observatory of Mallorca (OAM) from where more than 3,000 asteroids have been discovered.

Other observing programmes stem from other initiatives such as the Gaia Follow Up Network for Solar System Objects (Gaia-FUN-SSO), which arose from the Data Processing and Analysis Consortium of the Gaia mission and is expected to discover a large number of asteroids. At this time, when GAIA's goals are not yet fully fulfilled, it has been reported that this probe has discovered more than 14,000 asteroids. Obviously its fine astrometry also allows to accurately compute asteroid orbits and also improve that of other previously known asteroids.

In order to find out which objects are the most dangerous among the NEO population, the category 'potentially hazardous asteroid' was defined. Potentially hazardous asteroids are those objects that have an orbit with a minimum orbital intersection distance with the Earth of less than about 7.5 million kilometres and, in addition, a diameter of more than about 150 km. The latter parameter is usually approximate, being inferred from the absolute magnitude of the object and the average reflective capacity of these objects. Only close approximations allow the use of large ground-based radio telescopes to probe the shapes and true dimensions of these dangerous objects, which are generally too distant, faint, imperceptible points of light, and embedded in the vastness of our planetary system.

As we saw in the previous chapter, the asteroid population in the near-Earth region is significantly large. At the end of December 2021, we know the orbits of 27,943 NEOs, of which 27,826 are asteroids and 117 are comets. Among them are 2,239 potentially hazardous asteroids (also known as PHAs) of which 160 are larger than one kilometre in diameter. The latter PHAs would be capable in an impact of wiping out life as we know it from the Earth's surface, but fortunately there is no danger in the short or long term. Nevertheless, programmes to track and discover new NEOs have an important, and not inconsiderable, observational bias. Tracking these objects from ground-based observatories means having unexplored regions of the sky where, depending on the season of the year and the position of the Sun, potentially dangerous objects escape us. Thanks to dynamical simulations over millions of years, and the study of the so-called cosmic-ray exposure ages (CREAs) (Eugster et al., 2006), we know that these bodies transit the near-Earth region for a few million years and then return to the main belt. Therefore, the NEA population on the astronomical scale is highly dynamic and changing, reflecting the objects that fall into resonances within the main belt as a consequence of the non-gravitational forces that I will describe below. These near-Earth asteroids are classified according to the values of the semimajor axis, the distance to perihelion and the distance to aphelion of their orbit (Fig. 5.4).

Discovering Near-Earth Asteroids

NEAs can be classified into several groups according to the value of the semi-major axis of their orbit (a) and the values of the aphelion (Q) and perihelion (q)

distances (Fig. 5.4). Thus, those whose orbit meets the conditions $a > 1.0$ AU and $q < 1.02$ AU are called Apollo and form the largest group: approximately 62% of the NEAs identified so far fall into this classification. These objects cross the Earth's orbit, as do the Aten-type NEAs, which are those asteroids whose semimajor axis is less than 1 AU, with distances to aphelion less than 1.0167 AU. The third group is the so-called Amor-type asteroids, which have $a > 1$ AU and distances to perihelion between 1.02 and 1.3 AU. The orbits of Amor-type asteroids therefore fall entirely outside the Earth's orbit, although they approach our planet when they are at perihelion. A final group are the IEOs, an acronym for Inner Earth Objects, whose orbit is at all times contained within the Earth's orbit. To date, only six NEAs have been identified that fall into this category.

Near-Earth objects include potentially hazardous asteroids, also known as PHAs (*potentially hazardous asteroids*). They are specifically defined as NEOs whose orbits have a minimum orbital intersection distance of 0.05 AU and which are larger than about 150 m in diameter. At the end of December 2021, we know of 2,239 PHAs and, among them, there are 160 with an estimated diameter of more than one kilometre. As shown in Fig. 5.4, there are 10,121 NEOs in Amor-type orbits, 15,517 in Apollo-type orbits, 2161 in Aten-type orbits, 27 in Atira-type orbits and 117 comets in the near-Earth region (NECs).

Despite the progress made, it seems obvious that monitoring programmes should be extended to cover the vast majority of objects with a diameter capable of causing danger on a local scale. A lower limit could be around 20 m in

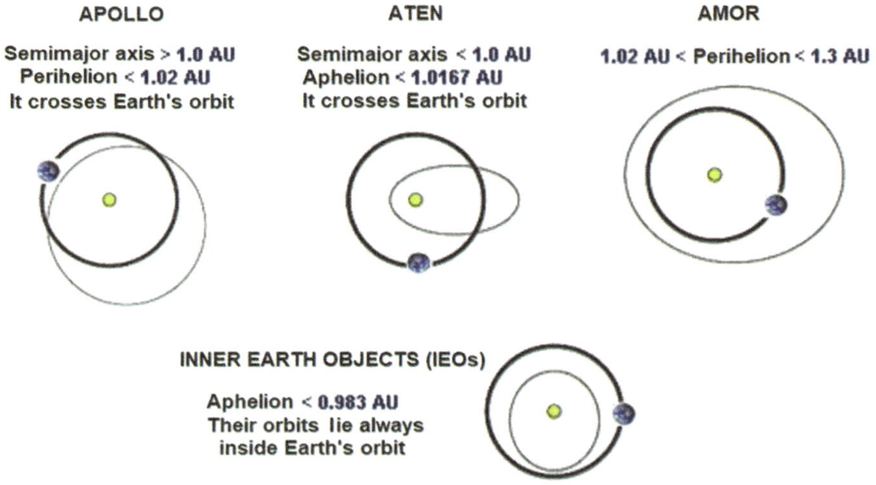

Fig. 5.4 Classification of NEAs according to the values of the semimajor axis, the distance to perihelion and the distance to aphelion of their orbit

diameter, the approximate dimensions of the Chelyabinsk event. This leads inexorably to the need to promote the development of space telescopes with infrared detectors fully dedicated to sky patrol. At this wavelength these asteroids are particularly bright because they radiate the heat they receive from the Sun, even on their faces not directly illuminated in the particular geometry of the observation.

Many of the NEOs of moderate diameter that are discovered are only recorded in small arcs of their heliocentric orbits, just at the dates of closest proximity to the Earth. In the rest of their orbit they are unobservable objects and, therefore, the orbits derived for most of them require extrapolating a small arc. This implies significant errors in the derivation of the parameters defining their orbits, the so-called orbital elements, and in the predicted date of their return. This is why many researchers support the construction of a space telescope dedicated to the search and tracking of these objects, sensitive to the infrared. Such a spacecraft could be strategically located in a favourable orbital geometry for the detection of the elusive NEOs and, from that privileged perspective, would be able to significantly increase astrometric observations of these elusive asteroids. Being able to track their movements over much longer arcs would significantly improve knowledge of their orbits.

Every time an asteroid approaches our planet, the race is on to find out more details using all available techniques. Using the largest radio telescopes, such as the Goldstone and, until recently, the Arecibo telescope, it has been possible to probe the shapes of these asteroids as they approach the Earth. In a simple way, the sounding procedure consists of sending out the radio signal and listening to the echo that the asteroid sends back to the radio telescope. This makes it possible to reconstruct the shape and structure of these rocky or metalloid objects, and even to delve into the main details of their surface, whether craters or large boulders.

5.3 Asteroids Subject to Capricious Forces

Although the tracking programmes are comprehensive in that they cover the celestial sphere from the ground, it has already been mentioned that there are unobservable regions which may contain elusive objects. Space telescopes are therefore already being designed to complete the search of the entire celestial sphere. The idea of the *Sentinel* mission, or other initiatives that aim to install space telescopes in space to detect asteroids while minimising the observational bias associated with the geometry in which the object approaches the Earth, therefore makes sense. There are still undiscovered objects of a few hundred kilometres in diameter, although most of the unexpected objects must be tens of kilometres across. The best example of an unexpected visitor, and a very recent one, was the Chelyabinsk meteorite fall on

15 February 2014, produced by an object around 20 m in diameter. Another historic encounter on 30 June 1908 was truly devastating and became known as the Tunguska (Siberia) event because of the Russian region it devastated. The blast wave generated by the fragmentation of a body with a diameter of about 30–50 m in diameter,flattened approximately 2,150 km^2 of Siberian taiga, literally sweeping away 80 million trees. Let's look at why these objects dangerously cross the Earth's orbit and why, every century, they can potentially produce these locally devastating processes.

The orbital dynamics of asteroids approaching the Earth-swept region are complex because they are subject to a number of non-gravitational forces. The so-called *Yarkovsky effect*, associated with the balance of the solar radiation reaching the surface of an asteroid and the part radiated preferentially in the infrared by the non-illuminated face, has a key influence on the dynamical evolution of asteroids (Bottke et al., 2002a). It is named after Ivan Yarkovsky (1844–1902), a Russian civil engineer who investigated the effects of radiation on small solar system objects. He proposed that the infrared radiation emitted by a body heated by the Sun also transfers momentum and heat to space. Such energy loss takes a particular toll on small asteroids and causes them to lose orbital energy, i.e. to decay progressively and slowly towards the Sun. It is undoubtedly a subtle force, arising from the asymmetry between the energy absorbed and radiated by an asteroid respectively from its illuminated and unilluminated faces (Chapman, 2010). It was not until the early twenty-first century that it was openly recognised that meteoroids, asteroidal fragments up to 10 m in diameter, move in their orbits until they reach resonances with the planets that push them into the near-Earth region by the Yarkovsky effect and not by impacts as previously thought. Indeed, this is the case to such an extent that encounters or collisions with small objects, which are a cause for concern as they cannot be considered when integrating their orbits, could play a secondary role.

There is another even weaker force that is known by the surnames of its discoverers: *Yarkovsky–O'Keefe–Radzievskii–Paddack* which, for obvious reasons, is simplified by its acronym as the *YORP effect*. Its discovery and physical definition were the work of several researchers. In addition to the aforementioned Yarkovsky, it was the Russian astronomer Vladimir V. Radzievskii (1911–2003) who applied the idea of considering these radiative effects in the rotation itself based on changes in albedo (the ratio between reflected and incident light). Finally, Stephen J. Paddack (1934–pres.) and John A. O'Keefe (1916–2000) realised that the shape of the objects under study must have been a key factor in these processes effectively altering the rotation period of asteroids.

The asteroids and comets present in the near-Earth region come from the main asteroid belt where, by the action of the aforementioned physical processes, their orbits are progressively shortened, falling into the so-called resonances with the planets (Jupiter, Saturn and Mars being the most influential) where they acquire orbits eccentric enough to deviate towards the near-Earth region. It has recently been discovered that this is a very dynamic system and that in these near-Earth regions asteroids will orbit for at most a few million years before returning to the main belt (Bottke et al., 2002b). Many of these asteroids entering the inner solar system are in fact piles of debris that resemble several weakly aggregated blocks.

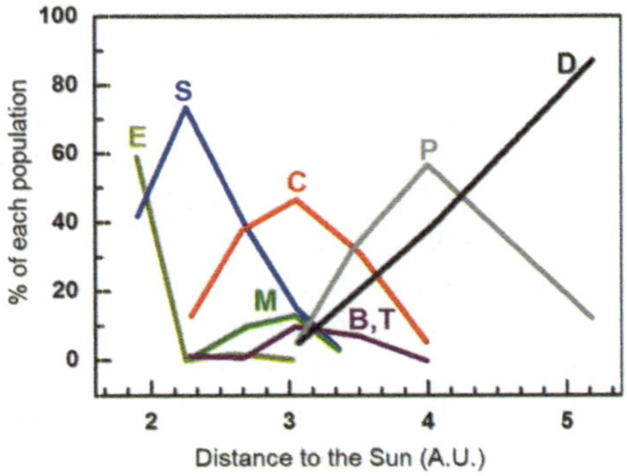

Fig. 5.5 Distribution of populations in the main asteroid belt. The classes of meteorites with which they can be associated are: E: enstatite asteroids; S: ordinary chondrites; M: metallics; B: carbonaceous type 2 Pallas; C: carbonaceous type 253 Mathilde; T, D and P: these are carbonaceous asteroids (Marco Polo-R team)

Not all of these asteroids in the main belt are of similar composition and structure. While rocky ones of a chondritic nature abound, others are essentially metallic and others are rocky-metal. We know of different groups or populations that are remotely characterised by their reflection spectra. The minerals that make up these asteroids have spectral bands and appearances similar to those of certain minerals. Thus, there is a population of asteroids rich in a silicate formed under reduced conditions called enstatite, which gives its name to the E group. Then there are those that show spectral bands characteristic of two other common silicates in ordinary chondrites called olivine and pyroxene. This class is known as S because of the curious shape of its reflection spectrum. Then there are the M-type asteroids which are metallic. Then there are asteroids of low reflectivity and generally free of pronounced absorption bands, which are carbonaceous in nature. These include spectral class B asteroids possibly associated with asteroids such as (2) Pallas and spectral class C asteroids such as (253) Mathilde. Other more minor groups of carbonaceous nature are the spectral classes T, D and P. The distribution of these populations in the main belt is shown in Fig. 5.5. It can be seen that this distribution is quite heterogeneous and is reminiscent of the processes that have been sculpting this region of our planetary system over the eons.

5.4 Spinning-Top Asteroids

Asteroids rotate about themselves, either around one axis or around several axes, by virtue of their intrinsic shape and complexity. This rotation is often referred to as spin, a term I will use from now on. One of the consequences of the *YORP effect* is that asteroids can have their spin accelerated depending on the radiation conditions and the structure of the object itself. This can have important consequences when asteroids are made up of faint piles of debris since, if their rotation speed increases or they are subjected to an external gravitational field, such as when they pass close to a planet, they can disintegrate (Trigo-Rodriguez et al., 2007).

It was Alan Harris (1996) who discovered that the internal structures of objects larger than 150 m had rotation periods longer than 2.2 h. The reason for this limit on the rotation period of an object is that these asteroids are mostly piles of debris, fragments resulting from colossal impacts between larger asteroids. In one of these rocky block systems, gravity causes them to be loosely stacked so that rapid rotation leads to their destruction by centrifugal force. It has therefore been found that the vast majority of asteroids of this size have such a limitation on their rotation period, since those that have exceeded it have suffered catastrophic fragmentations. This is a magnificent demonstration that there are very few monolithic objects in the near-Earth region.

Among these potentially dangerous, rapidly rotating asteroids, there are some really big ones. One example is the Amor-type asteroid (52,768) 1998 OR2 which, at 2 kms in diameter, is one of the largest and makes frequent approaches to our planet in an orbit with a period of only 3.68 years. Such highly eccentric NEAs move very fast in our images and exemplify the imperative need to use radio telescopes to reconstruct their silhouettes (Fig. 5.6).

Other unique asteroids provide valuable lessons, such as the so-called super-fast rotators (SFRs) among which we find, to give a couple of examples, 2005 UW163 (Chang et al., 2014) and 2011 XA3. To give us an idea of how they behave, the latter object is about 200 m in diameter and rotates on its axis in less than 44 min (Urakawa et al., 2014). We still do not know the nature of many of these asteroids because their properties are variable due to the intrinsic complexity of obtaining details about their

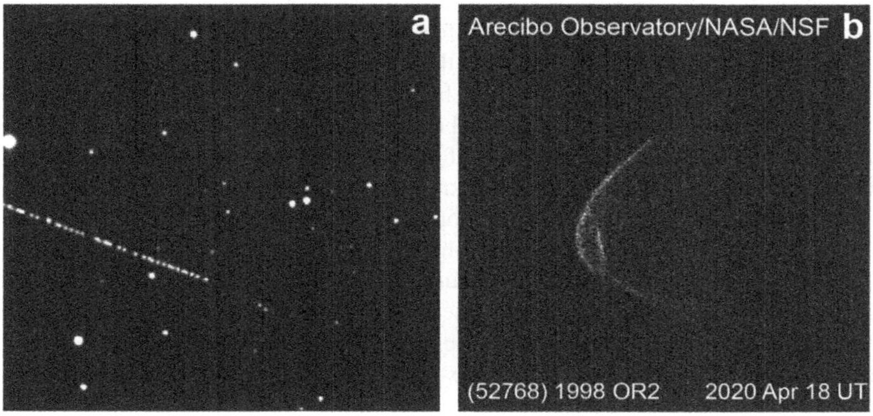

Fig. 5.6 The potentially hazardous Amor group asteroid (52,768) 1998 OR2. **a** Author's image from 25 April 2020 showing a motion of about 10 arc minutes in only 30 min of exposure from the Observatori del Montseny. **b** Radar image of the asteroid obtained a few days earlier by the Arecibo radio telescope in Puerto Rico

internal structure. In addition, their crazy rotation makes it difficult to interpret the reflection spectra that can be obtained. Here are some clues about some of these objects whose reflective behaviour has been shown to be surprising.

5.5 The Tidal Effect in Near-Earth Passages

Some interesting properties of near-Earth asteroids lie in their spectral behaviour. For decades astronomers have been greatly intrigued by the fact that the reflective properties of main-belt asteroids did not look much like those of near-Earth asteroids and not at all like the meteorites we study in our laboratories. It was from the Apollo missions, particularly those brave astronauts who spent many hours of extravehicular activity (EVA) in collecting almost 400 kg of lunar rocks, that a detailed study of the mineralogical composition of the Moon could be undertaken in the clean rooms that NASA set up for the occasion. It became clear that the lunar rocks, and those that must have existed on the surface of many asteroids, had undergone an effect known as space *weathering*, in which cosmic radiation, solar wind particles and micrometeorites had obscured their reflective properties. In simplified terms, we could say that the asteroids were 'discoloured' by these processes. Such findings led to the proposal that in the near-Earth flybys the regolith-covered surfaces must have moved tidally and regenerated, changing the asteroid's spectrum (Nesvorný et al., 2005). Still, the enormous diversity found could not be explained.

Subsequently, Richard Binzel's team at the Massachusetts Institute of Technology (MIT) undertook a detailed study of the reflective properties of NEAs and representative meteorites (so-called chondrites). It should be noted that most NEAs fall into the category of ordinary chondrites, and in their close flybys they can have their surface removed by gravitational tidal effects (Binzel et al., 2010). Well, the research team found that some NEAs resembled these chondrites, but others resembled their possible progenitors located in the main belt. In any case, they found diverse objects between the two extremes and proposed that there must be very dynamic processes capable of regenerating the surfaces of certain asteroids. Thanks to the work of Pierre Vernazza and collaborators, we know that asteroids undergo such processes on timescales typically less than a million years (Vernazza et al., 2009).

At that time, we had already found meteoroids associated with NEAs, such as the 2002 NY40 complex (Trigo-Rodríguez et al., 2007). We proposed then that the gravitational tug effect at close approaches to the Earth or other planets could play an important role in detaching noticeable rocks from the surface of these objects, or even fragmenting them if the approach was extreme. The idea was quite revolutionary, as the orbits of few meteorites were known at the time and it was mostly established that they came directly from the main belt, driven by resonances. Would there after all be other mechanisms for transporting meteorites into Earth's orbit?

5.6 Dating the Stay of Meteorites in Space

The degree of sophistication achieved in the study of isotopes contained in meteorites may be surprising. The amount of information that can be extracted about the sojourn of meteorites in interplanetary space is particularly surprising and useful to confirm the dynamic transport scenarios outlined above. A generally little-known aspect is that we can know even more than the exact age of formation of meteorites from the chronometers provided by their isotopes. Before the meteorite reaches the Earth's surface it will have travelled through space as a meteoroid. There is a type of radioactivity that is induced by exposure of the meteoroid to galactic cosmic rays and energetic particles that are produced in solar flares and transported by solar wind to all corners of our planetary system. Some particles are literally implanted in meteoroids, others produce traces in the materials discernible when viewing those materials through *transmission electron microscopy (TEM)*, while in the most energetic cases the particles hit the atoms of meteoritic minerals with such intensity that they create stable or radioactive discrete nuclei that are described as *cosmogenic*. Based on the analysis of the abundances of these cosmogenic nuclei, produced by cosmic-ray bombardment, it is possible to infer for each meteorite the so-called cosmic-ray exposure ages (CREAs).

Physical models that allow modelling of the exposure of meteoritic materials to cosmic rays provide information on how long these materials have travelled through interplanetary space since leaving their parent asteroid. They also provide information on the dimensions of the meteoroid, the percentage of mass lost in atmospheric ablation, and the depth at which the materials that make up the surviving meteorites were found.

Today we know the CREAs of hundreds of meteorites (see Fig. 5.7). These studies provide us with a valuable estimate of the transfer times of meteorites from the time they leave their parent asteroids and reach the Earth. Ordinary and enstatite chondrites in particular have the longest transfer times, in the order of tens of millions of years. This time is much shorter for carbonaceous chondrites, which are also much rarer. This indicates that, depending on various processes affecting these rocks during their stay in the interplanetary medium, important biases occur (Eugster et al., 2006). On the one hand, ordinary chondrites are much more consistent and survive fragmentation processes that expose them to cosmic rays (Eugster et al., 2006). In fact, we know that collisional processing of main-belt asteroids has been very important since the formation of the solar system and many asteroids have lost a very significant part of their mass (Beitz et al., 2016). This is evidence that very few asteroids have remained unaffected by catastrophic collisions.

Thus, the distribution of CREAs in the different chondrite groups indicates that most of them originate from collisions in the main asteroid belt, which occurred between 5 and 50 Ma (Eugster et al., 2006). This is true even for most groups of carbonaceous chondrites, although the most primitive ones associated with the CI and CM groups that we suspect could be comet-derived have CREAs of less than

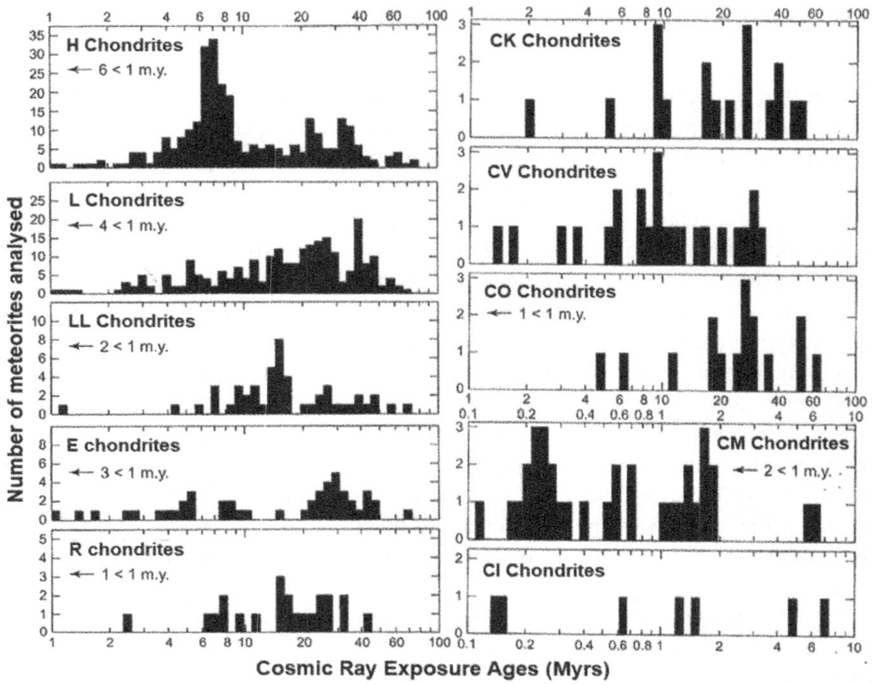

Fig. 5.7 Diagram of cosmic-ray exposure ages (CREAs) for ordinary and enstatite chondrites (left) and for carbonaceous chondrites (right). The arrows below the name of each group indicate the number of meteorites with CREAs less than one million years old that are not shown in the diagram. Figure adapted from Eugster et al. (2006) with permission of the authors

about 10 Ma (see Fig. 5.7). Therefore, there must be mechanisms that destroy these more fragile meteoroids and, possibly, their parent bodies on these timescales.

It has also been proposed that close approaches can regenerate the surfaces of NEAs by removing the regolith surfaces affected by space weathering. The gravitational tidal effect should reorient the regolith on the surface of such objects and make their reflective properties more similar to the meteorites found in terrestrial collections (Binzel et al., 2010).

Quantifying the Impact Hazard: The Torino Scale Understanding the impact hazard is not a trivial matter. Scientists wish to explain this key topic to the population in an understandable way. For precisely that reason the Torino Impact Hazard Scale was created as a consequence of an international meeting on the matter that took place in the Italian city of Turin in 1999. The scale in question assigns values from 0–10 to the risk posed by asteroids and comets. Objects in close approaches with no risk may have values between 0 and 1,

Table 5.3 The Torino Impact Hazard Scale, with the main criteria chosen to popularize the topic, as it was published in 2000

Main consequences	Torino impact hazard scale	Main explanation
Events having no consequences	0	The likelihood of a collision is zero, or well below the chance that a random object of the same size hits the Earth within the next few decades. This degree also applies to any small object that, in case of a collision, is unlikely to reach the Earth surface intact
Events meriting monitoring	1	The chance of a direct collision is extremely unlikely, about the chance that a random object of the same size hits the Earth within the next few decades
Events meriting concern	2	A somewhat close, but not unusual encounter. Collision is very unlikely
	3	A close encounter, with 1% or greater chance of a collision capable of causing localized destruction
	4	A close encounter, with 1% or greater chance of a collision capable of causing regional devastation
Threatening events	5	A close encounter, with a significant threat of a collision capable of causing regional devastation
	6	A close encounter, with a significant threat of a collision capable of causing a global catastrophe
	7	A close encounter, with an extremely significant threat of a collision capable of causing a global catastrophe
Certain collisions	8	A collision capable of causing localized destruction. Such events occur somewhere on Earth between once per 50 years and once per 1,000 years
	9	A collision capable of causing regional devastation. Such events occur somewhere on Earth between once per 1,000 years and once per 100,000 years

(continued)

Table 5.3 (continued)

Main consequences	Torino impact hazard scale	Main explanation
	10	A collision capable of causing a global climatic catastrophe. Such events occur somewhere on Earth between once per 100,000 years, or less often

while a value of 10 indicates a global catastrophe produced by a certain impact from a km-sized body or even larger (see Table 5.3). In general, most events fall into the categories 0 or 1.

It should be noted that the Torino scale was not automatically accepted. However, in November 2000 the global media were again producing scary news items about the close approach of a small Aten-type asteroid 2000 SG344. This body of about 40 ± 25 m in diameter had potential to produce a new Tunguska-sized impact event. About one month after its discovery on 29 September 2000, once several pre-discovery observations allowed the astronomers to improve the accuracy of the close approach path, the impact was almost ruled out. At that time, Paul Chodas from the Near-Earth Object Program Office at NASA Jet Propulsion Laboratory announced a 1 in 500 probability of impact with our planet on 21 September 2030. Astronomers learned that a premature post-discovery announcement of any event presumed to occur many years ahead can make matters worse, producing unnecessary alarm that in the end results in a loss of public confidence with the scientists. The assessment of the Torino scale synthesized in Table 5.3 is a non-trivial issue because there are many factors at play in each event. In particular, a very close encounter does not necessarily mean that the impact is produced, as exemplified by the asteroid (367,943) Duende encounter passing at 27,700 km from the Earth's surface. Often, an asteroid needs first to cross over a really narrow region nearby the Earth, called a *keyhole*, in which the gravitational effect of the encounter puts the asteroid on a trajectory which finally produces the impact at a later encounter.

5.7 The Spaceguard Asteroid Discovery Milestone

A key milestone in our progress detecting NEAs was the publication of the *Spaceguard Survey Report* (Morrison, 1992). It was written after a US Congress request and led to a clear goal: NASA should locate 90% of NEAs larger than 1 km within one decade. Such a milestone was in fact the start of a race to discover NEAs, and it is often referred to as the *Spaceguard Goal*. Looking at the progress in asteroid discoveries in perspective, such a report became an inflection point to the development of search initiatives, because the US government was funding telescopic surveys that received money through NASA. All these surveys were working in the framework of the so-called *Spaceguard Project*.

Technological innovation in astronomy is well exemplified in this field because these surveys used modern digital detectors called charge-coupled devices (CCDs) to increase the efficiency in asteroid detection. Actually, as mentioned, it was Tom Gehrels in the Spacewatch survey who first used a CCD to discover a NEO in 1989. Such a discovery motivated the researchers to use the capacities of the new available digital cameras to increase the capacity of discovering tenuous, fast-moving NEAs.

Obviously, to accomplish the Spaceguard Goal without CCDs had been really difficult, and even using them it took the surveys until the end of 2005 to achieve 75% of that milestone for km-sized asteroids. Fortunately, the search effort did not cease and new initiatives were created so the milestone was achieved in 2017. Once 90% of the km-sized NEAs were discovered, the NEO Program focused on finding 90% of the NEO population larger than 140 m. The discoveries have increased almost exponentially since 1992, as can be seen in Fig. 5.8.

To achieve the new discovery goal, asteroid surveys established searches in sky regions with moderate inclinations, in other words not so concentrated in bodies moving near the ecliptic plane, a decision that was a step ahead and ended in the discovery of other interesting objects. Highly eccentric dormant comets are certainly among them, but also interstellar visitors such as 1I/'Ouamuamua. This fascinating object that had travelled interstellar space for millions of years was discovered on

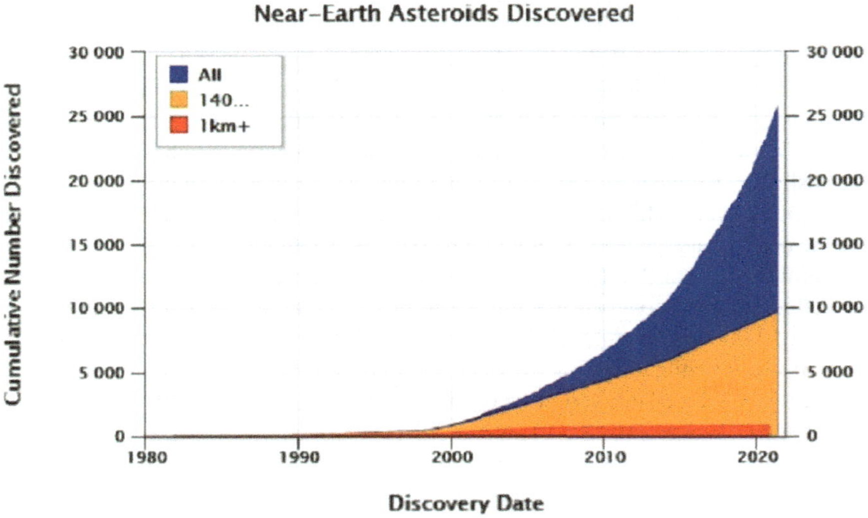

Fig. 5.8 CNEOS/JPL cumulative graph showing the number of near-Earth objects discovered since 1980 up to 18 May 2021, following the systematic searches promoted by NASA under the US Congress request. Red is for km-sized objects, yellow for asteroids larger than 140 m and blue for all known NEAs

19 October 2017 by Robert Weryk, using the Pan-STARRS telescope at Haleakalā Observatory, Hawaii. Despite the speculative theories about being an extraterrestrial probe, this object had spectral properties reflecting its natural origin after a long journey in the interstellar medium (Meech et al., 2017). Very elongated asteroids can be produced during the early phases of planetary evolution and gravitationally scattered by giant planets, as we know happened in our own solar system (Walsh and Morbidelli, 2011).

Nowadays there are different initiatives studying possible impacts. The Center for Near-Earth Object Studies has created and automated a collision monitoring system called *Sentry*. This powerful system scans an updated asteroid catalogue searching for chances of future impact with Earth of any object over one century timescale. This initiative provides an online table ranking the objects with a cumulative probability of impact (see the online resources). The increasing discovery of small asteroids, from a few metres up to tens of metres in diameter, is also leading us to detect asteroids on direct collision routes with the Earth, like 2008 TC3 or 2018 LA (Jenniskens et al., 2009; 2021).

5.8 The Current Danger of Impact by Small Asteroids

By compiling all the available information on both close encounters with asteroids and the large meteoric bolides produced to date, it is possible to pinpoint the statistical frequency at which our planet receives notable impacts with metre-sized projectiles. Brown et al. (2013) established that the frequency of these collisions varied with the energy of these impacts following the data shown in Fig. 5.9. Thus, these data have been taken by the scientific community as a reference for these estimates, although other researchers using rock impact analysis techniques against the Moon (Ortiz et al. 2006) have already warned that the real impact rate was much higher than that established by Brown et al. (2013). In fact, following the analysis of the analysis of the Chelyabinsk event, which took place in Russia in February 2012, that team concluded that the impact rate for objects larger than 20 m in diameter was increased by an order of magnitude. From this frequency depending on the diameter of the projectile we can deduce that an impact event with an asteroid of about 20 m in diameter and a mass of about 12,000 tonnes—i.e. of the dimensions of Chelyabinsk—would occur every 50 years. Meanwhile, an impact with an asteroid of about 50 m in diameter, such as the one that produced the Tunguska event in 1908, would occur every 500 years (see next chapter).

Although statistically the impacts appear well spaced, there could be additional sources of small asteroids and large meteorite-producing meteoroids, not directly sent from the main belt by the dynamical mechanism of planetary resonances. In fact, some primitive carbonaceous chondrites have CREAs that are a few million years old or even less than 500,000 years old. They may have been produced by the recent fragmentation of evolved comets, but they could also come from other transport mechanisms to Earth. For example, many NEOs have a structure called *rubble*

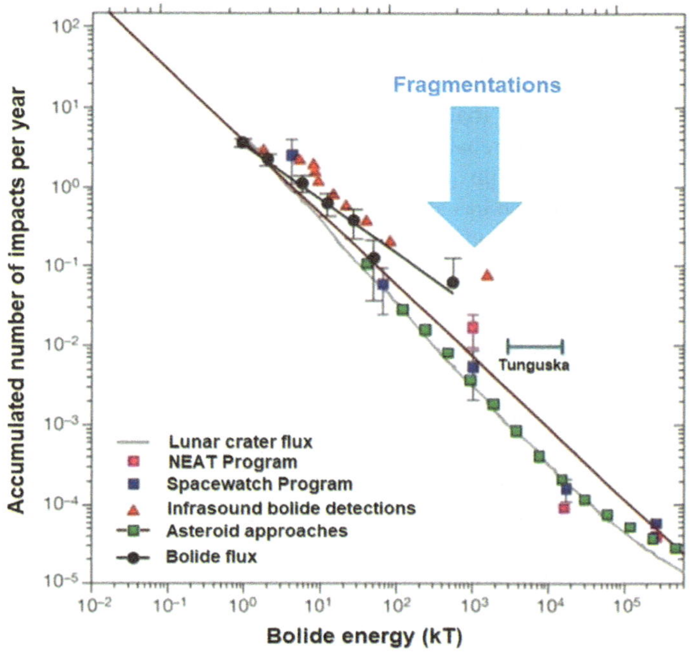

Fig. 5.9 Cumulative frequency of Earth impacts as a function of impact energy. The various lines indicate different methods of flux estimation based on bolides or asteroid encounters (adapted from Brown et al., 2013)

piles, so it is possible that in their approaches to the terrestrial planets, sufficiently large rocks could have been detached by the gravitational tidal effect or even fragmented into their constituent blocks (Trigo-Rodríguez and Williams, 2017). Current statistics point out that the Earth is impacted by about 40 m-scale objects every year (Devillepoix et al., 2019).

The analysis of the Earth impact flux of objects in the NEO region is of paramount importance, mainly because of the catastrophic consequences that these events could have. We know that the larger the size of the colliding object, the much smaller its probability of impacting the Earth, because there are far fewer large objects than small ones, roughly following a power law. We think that a devastating impact, with biospheric consequences capable of destroying our civilisation, would occur for an asteroid with a diameter of about one kilometre. The latter would hit the Earth at an interval of about a million years. Much larger objects are, fortunately, much rarer and impact events with an asteroid with a diameter of about 10 km, such as the one that produced the Chicxulub crater, usually occur every 100 million years.

The impact flux associated with small asteroids can be estimated using various techniques. Infrasound stations that were set up to monitor nuclear explosions have in the last decade been used to detect some of the most energetic events recorded, often appearing at the most remote locations imaginable. For more frequent small bodies, from centimetre- to metre-sized meteoroids, a better understanding can be indirectly achieved from the study of the bright bolides that they produce, once recorded by imaging networks (Moreno-Ibáñez et al., 2015). Thus, the analysis of superbolides recorded by the Spanish Fireball and Meteorite Research Network (SPMN) shows that for smaller objects this flux should also be corrected upwards (Trigo-Rodríguez and Madiedo, 2013), which has also been confirmed by the detection in September 2013 of the most luminous lunar impact ever recorded (Madiedo et al. 2014).

5.9 What Lessons Should We Learn?

In recent decades we have become increasingly aware of the impact hazard (Chapman and Morrison, 1994). The Torino scale was designed in 1999 to communicate to the public the risk associated with a possible collision of an asteroid or comet with the Earth. This scale, with integer values from 0 to 10, takes into account the predicted impact energy of the event, as well as the impact probability of the event (Binzel, 2000). Furthermore, in the last decade we have conveyed to the public our concern and the relevance of undertaking detailed studies of the physical properties of asteroids and comets. It is within this constructive international framework that *Asteroid Day*, an annual global event celebrating the anniversary of the Tunguska event on 30 June 1908, was created. The United Nations has recognised the day on a global scale to help raise awareness of our enormous fragility in the face of a major impact (Trigo-Rodriguez et al., 2017).

Asteroids and comets have been the target of several space missions in the last decade. They are essential to understand the very nature of these bodies and to learn how to cope with their onslaught. However, there is no doubt that this is not enough: more needs to be done to develop the aerospace technology needed to deal with and mitigate encounters with these dangerous space wanderers.

In this sense, we are taking a major step forward with the Double Asteroid Redirection Test (DART) mission, an interdisciplinary collaborative test involving American and European researchers to develop and implement the first kinetic impactor to deflect an asteroid. DART will be the first demonstration of the kinetic (non-explosive) impact technique to change the motion of an asteroid in space. The asteroid chosen for such a scientific and technological milestone is the satellite of the binary asteroid (65,803) Didymos. While Didymos' primary body is about 780 m in diameter, its satellite called Dimorphos is about 160 m, in the typical size range that we might face in a future medium-term impact threat. This pioneering kinetic impactor test will be followed by the dedicated Hera ESA mission that will explore the binary asteroid, quantifying the consequences of DART crater excavation in the small satellite (Michel et al., 2018).

The race to learn about asteroids continues. The European Union has therefore promoted initiatives for monitoring and mitigating such encounters under the H2020 programme. There are however aspects of the intrinsic hazard associated with asteroid and comet impacts with the Earth that are still unquantified and worthy of concern. This is precisely why we need to apply a new technique to learn from the mechanical properties of meteorites and asteroids (Tanbakouei et al., 2019). An encounter with one of these objects of several hundred metres that has not yet been discovered could be both unexpected and catastrophic for humanity. Let us therefore view the encounter of these asteroids with curiosity, but also as a disturbing warning. We are gradually creating space telescopes suitable for tracking them, although we will need more specialised ones, especially those skilled in infrared searches where they are more easily detectable, to increase the time window for discovering these objects of tens of metres. Being able to catalogue them and determine their orbits in good time will prevent them from endangering our very existence.

Dart: Our Ability to Deflect Asteroids

In 2022, humanity will carry out its first on-site test to implement a direct palliative technique: the deflection of an asteroid. The mission to make this substantial step is called the *Double Asteroid Redirection Test*, or just *DART*, from the acronym of this first planetary defense mission itself. This initiative aims to put into practice our ability to deflect asteroids through a technique that we could consider simpler: the kinetic impactor. The binary asteroid 65,803 Didymos has been chosen, specifically not its central primary body of 780 m but its Dimorphos satellite, about 160 m in diameter, for this historical test of our ability to deflect asteroids. DART, designed for its kamikaze operation, consists of a main spacecraft equipped with modern instrumentation with its antennas, thrusters and other instrumentations with dimensions of $2.1 \times 1.9 \times 2.6$ m and with two long solar panels that extend 18.9 m from end to end. they will impact the moonlet at high speed. With its impact against Dimorphos at a speed of 6.6 km/s it is intended to demonstrate the technique of the kinetic impactor to test the efficiency of momentum transfer by a projectile direct hit.

The DART mission has several fundamental requirements to become a successful mission. First, the DART probe is expected to hit Dimorphos less than a year after taking off on Nov. Specifically, the impact against Dimorphos is expected to be between September 25 and October 2, 2022, when it is about 11 million km from Earth, causing at least a change of 73 s in its rotation period around the primary asteroid. Secondly, we have the challenge of measuring the change in the binary asteroid period with an uncertainty of less than 7 s and quantifying the efficiency of the transfer at the moment as a result of the impact, a parameter known as Beta (β). Obviously, the Hera probe of the European Space Agency (ESA) has the goal of visiting the Didymos system after the crater excavation by DART in order to characterize the effects resulting from the impact. The data necessary to meet these requirements will be obtained and analyzed by the DART Research team in which I am honored to participate as part of the science team. It is undoubtedly a much-needed mission to begin to have a clear idea of the effectiveness of this method in order to alleviate encounters with small asteroids, those of the dimensions of the object that caused the Tunguska event on June 30, 1908.

Questions for Chapter 5

Q.1. Imagine that tomorrow is discovered a body exhibiting an albedo of 0.1 and an absolute magnitude of H = 27. (a) Many media reports start to announce the end of the world—are they correct? Use Table 5.4 to infer the asteroid's approximate diameter. (b) Imagine that there was an error of transcription and the asteroid's absolute magnitude was H = 17—how does it change the outcome? [Solution: (a) The asteroid's diameter is about 17 m, so a Chelyabinsk-like fireball is expected to be produced with local consequences; certainly it won't be the end of our civilization. (b) The scenario changes completely as H = 17 with the same albedo corresponds to 1.7 km, so global devastation could be on the way.]

Q.2. After a careful astrometric reconstruction of the orbital path, the asteroid explained in the previous question is found to have an almost direct route of impact with the Earth. The reconstruction of the encounter geometry indicates a presumed impact velocity of 14 km/s. Calculate the energy of the impact, assuming a spherical shape for the body and a carbonaceous chondrite composition from its relatively dark albedo. Use 2 g/cm^3 as the average density of carbonaceous chondrites. [Solution: From the approximate radius of the body given in Q.1 we can compute its volume. Then, we multiply the density by the volume to infer the pre-atmospheric mass: 5,145 metric tonnes. Be careful to use the same units. You can apply the law of propagation of errors to the formulae used in order to get the range of error.]

Q.3. Let's consider that the same asteroid is estimated to experience an impact over your country with a 3% chance. Sensationalist newspapers are first in alarming people with scary news stories without a real scientific basis. How will you use the Torino scale (see Table 5.3) to revert the effects of such bad journalist practice? [Solution: The estimated size of 17 m is somewhat similar to the Chelyabinsk fireball event. You can exemplify the presumed effects from that previous meteorite fall. Try to write a short report informing people about the most likely source of hazard (e.g. stay inside your home, preferably underground, to avoid exposure to the radiation produced by the bright flare in areas nearby the fall, and stay away from windows and doors to escape from the effects of the shock wave coupling with the ground that may produce seismic effects). On the Torino scale, such a sizeable body with a 3% chance of impact corresponds to level 3. Even when the asteroid impact is certain, it does not merit increasing the scale of risk because the asteroid is expected to break apart producing a meteorite fall.]

Q.4. Some asteroids could hide their real nature, so it is important to study them carefully before developing any mitigation strategy. A good example is asteroid Lutetia, which was visited by ESA Rosetta spacecraft. Its surface mineralogy was found to produce a reflectance spectrum like the one exhibited by enstatite chondrites, but other characteristics suggested a surface composition compatible with

Table 5.4 This table shows the approximate asteroid size (diameter in km) as a function of absolute magnitude H (left column) and visual geometric albedo (upper row). See Question 1 for a discussion on the limitations of these data (source: CNEOS/JPL)

albedo H	0.30	0.25	0.20	0.15	0.10	0.05
30.0	0.0025	0.0027	0.0030	0.0035	0.0043	0.0060
29.5	0.0031	0.0034	0.0038	0.0044	0.0054	0.0076
29.0	0.0039	0.0043	0.0048	0.0055	0.0068	0.0096
28.5	0.0049	0.0054	0.0060	0.0069	0.0085	0.012
28.0	0.0062	0.0068	0.0076	0.0087	0.011	0.015
27.5	0.0078	0.0085	0.0095	0.011	0.013	0.019
27.0	0.0098	0.011	0.012	0.014	0.017	0.024
26.5	0.012	0.014	0.015	0.017	0.021	0.030
26.0	0.016	0.017	0.019	0.022	0.027	0.038
25.5	0.020	0.021	0.024	0.028	0.034	0.048
25.0	0.025	0.027	0.030	0.035	0.043	0.060
24.5	0.031	0.034	0.038	0.044	0.054	0.076
24.0	0.039	0.043	0.048	0.055	0.068	0.096
23.5	0.049	0.054	0.060	0.069	0.085	0.12
23.0	0.062	0.068	0.076	0.087	0.11	0.15
22.5	0.078	0.085	0.095	0.11	0.13	0.19
22.0	0.098	0.11	0.12	0.14	0.17	0.24
21.5	0.12	0.14	0.15	0.17	0.21	0.30
21.0	0.16	0.17	0.19	0.22	0.27	0.38
20.5	0.20	0.21	0.24	0.28	0.34	0.48
20.0	0.25	0.27	0.30	0.35	0.43	0.60
19.5	0.31	0.34	0.38	0.44	0.54	0.76
19.0	0.39	0.43	0.48	0.55	0.68	0.96
18.5	0.49	0.54	0.60	0.69	0.85	1.2
18.0	0.62	0.68	0.76	0.87	1.1	1.5
17.5	0.78	0.85	0.95	1.1	1.3	1.9
17.0	0.98	1.1	1.2	1.4	1.7	2.4
16.5	1.2	1.4	1.5	1.7	2.1	3.0
16.0	1.6	1.7	1.9	2.2	2.7	3.8
15.5	2.0	2.1	2.4	2.8	3.4	4.8
15.0	2.5	2.7	3.0	3.5	4.3	6.0
14.5	3.1	3.4	3.8	4.4	5.4	7.6
14.0	3.9	4.3	4.8	5.5	6.8	9.6
13.5	4.9	5.4	6.0	6.9	8.5	12
13.0	6.2	6.8	7.6	8.7	11	15
12.5	7.8	8.5	9.5	11	13	19
12.0	9.8	11	12	14	17	24
11.5	12	14	15	17	21	30
11.0	16	17	19	22	27	38
10.5	20	21	24	28	34	48
10.0	25	27	30	35	43	60

primitive carbonaceous chondrites. By comparing its reflectance with the properties of carbonaceous chondrites from the Antarctic collection, we discovered a possible connection with the metal-rich CH3 chondrite called Pecora Escarpment 91,467 (Moyano et al., 2016). In fact, we can also infer the presence of metal from its bulk density. Taking into account the mass of 1.7×10^{18} kg and that the volume model inferred from Rosetta observations is about 5×10^{14} m^3, calculate its bulk density and try to find a way to increase naturally its bulk density. [Solution: The asteroid seems to be associated with carbonaceous chondrites, but its bulk density is clearly higher than is typical for these bodies. Taking into account that it is quite possibly macroporous as a consequence of impacts, the only way to increase its density is by the introduction of metal. The Lutetia asteroid should have a significant amount of metal grains in its interior.]

Online Resources

Asteroid Day: https://asteroidday.org/

Center for NEOs Studies (CNEOS): https://cneos.jpl.nasa.gov/

Near Earth Objects – Dynamic Site: https://newton.spacedys.com/neodys/

Sentry, Earth Impact Monitoring: https://cneos.jpl.nasa.gov/sentry/

Spacewatch Programme: http://spacewatch.lpl.arizona.edu/

LINEAR Programme: http://www.ll.mit.edu/mission/space/linear/

Near Earth Objects ESA Coordination Centre: https://www.esa.int/Safety_Security/Near-Earth_Object_Coordination_Centre

Near-Earth Objects Dynamic Site: https://newton.spacedys.com/neodys/

References

Beitz E., J. Blum, M.G. Parisi and J.M. Trigo-Rodríguez (2016) The collisional evolution of undifferentiated asteroids and the formation of chondritic meteoroids. *Astrophysical Journal* 824, A12, 29 pp.

Binzel, R.P. (2000) The Torino Impact Hazard Scale. *Planetary and Space Science* 48, 297–303.

Binzel R.P. et al. (2010) Earth encounters as the origin of fresh surfaces on near-Earth asteroids. *Nature* 463, 331–334.

Bland P. and N.A Artemieva (2003) Efficient disruption of small asteroids by Earth's atmosphere. *Nature* 424, 288–291.

Bottke W.F. et al. (2000). Understanding the distribution of Near-Earth Asteroids. *Science* 288, 2190–2194.

Bottke W.F. et al. (2002a) The effect of Yarkovsky thermal forces on the dynamical evolution of asteroids and meteoroids. In *Asteroids III*. by W.F. Bottke et al. (eds.), University of Arizona Press, Tucson, USA, pp. 395–408.

Bottke W.F. et al. (2002b) Debiased orbital and absolute magnitude distribution of the Near-Earth Objects. *Icarus* 156, 399–433.

Brown P. et al. (2002) The flux of small near-Earth objects colliding with the Earth. *Nature*, 420, 294–296.

Brown P. et al. (2013) A 500-kiloton airburst over Chelyabinsk and an enhanced hazard from small impactors. *Nature* 503, 238–241.

Chang C.-K. et al. (2014) A new large super-fast rotator: (335433) 2005 UW163. *Astrophysical Journal Letters* 791, L35, 6 pp.

Chapman, C.R. and D. Morrison (1994) Impacts on the Earth by asteroids and comets: assessing the hazard. *Nature* 367, 33–40.

Chapman C.R. (2010) Stripped on passing by Earth. *Nature* 463, 305–306.

Devillepoix H.A.R. et al. (2019) Observation of metre-scale impactors by the Desert Fireball Network. *Monthly Notices of the Royal Astronomical Society* 483, 5166–5178.

Eugster O. et al. (2006) Irradiation records, cosmic-ray exposure ages, and transfer times of meteorites. In *Meteorites and the Early Solar System II*, D. Lauretta, H.Y. McSween (eds.), University of Arizona Press, Tucson, USA, pp. 829–851.

Harris A.W. (1996) The rotation rates of very small asteroids: Evidence for 'rubble pile' structure. *Lunar and Planetary Science*, 27, 493.

Hills, J.G. and M.P. Goda (1993) The fragmentation of small asteroids in the atmosphere. *The Astronomical Journal* 105, 1114–1144.

Hsieh H. and D. Jewitt (2006) A population of comets in the Main Asteroid Belt. *Science* 312, 561–563.

Jenniskens P. et al. (2009) The impact and recovery of asteroid 2008 TC3. *Nature* 458, 485–488.

Jenniskens P. et al. (2021) The impact and recovery of asteroid 2018 LA. *Meteoritics & Planetary Science*, https://doi.org/https://doi.org/10.1111/maps.13653.

Jewitt D. (2008) Kuiper Belt and Comets: An observational perspective. In *Trans-Neptunian Objects and Comets*, Swiss Society for Astrophysics and Astronomy, Springer-Verlag, Heidelberg, Germany, pp. 1–78.

Johnson T.V. and F.P. Fanale (1973) Optical properties of carbonaceous chondrites and their relationship to asteroids. *Journal of Geophysical Research* 78, 8507–8518.

Madiedo J.M. et al. (2014) A large lunar impact blast on September 11th 2013. *Monthly Notices of the Royal Astronomical Society* 439, 2364–2369.

Marciniak A. et al. (2018) Photometric survey, modelling, and scaling of long-period and low-amplitude asteroids. *Astronomy & Astrophysics* 610, A7.

Meech K. et al. (2017) A brief visit from a red and extremely elongated interstellar asteroid. *Nature* 552, 378–381.

Michel P., M. Kueppers, H. Sierks et al. (2018) European component of the AIDA mission to a binary asteroid: Characterization and interpretation of the impact of the DART mission. *Advances in Space Research* 62, 2261–2272.

Morbidelli A., W. F. Bottke Jr., C. Froeschlé and P. Michel (2002) Origin and evolution of Near-earth Objects. In *Asteroids III* by W.F. Bottke, A. Cellino, P. Paolicchi and R.P. Binzel (eds.), University of Arizona Press, Tucson, USA, pp. 409–422.

Moreno-Ibáñez M., M. Gritsevich and J.M. Trigo-Rodríguez (2015) New methodology to determine the terminal height of a fireball. *Icarus* 250, 544–552.

Morrison D. (1992) *The Spaceguard survey*. Report of the NASA International Near-Earth-Object Detection Workshop.

Moyano-Cambero C.E., J.M. Trigo-Rodríguez, J. Llorca, S. Fornasier, M.A. Barucci and A. Rimola (2016) A plausible link between the asteroid 21 Lutetia and CH carbonaceous chondrites. *Meteoritics & Planetary Science* 51, 1795–1812.

Nakamura E. et al. (2012) Space environment of an asteroid preserved on micrograins returned by the Hayabusa spacecraft. *Proceedings of the National Academy of Sciences*, doi: pnas.1116236109.

Nesvorný D. et al. (2002a) The recent breakup of an asteroid in the main-belt region. *Nature* 417, 720–722.

Nesvorný D. et al. (2002b) Regular and chaotic dynamics in the Mean-motion resonances: Implications for the structure and evolution of the asteroid belt. In *Asteroids III* by W.F. Bottke, A. Cellino, P. Paolicchi and R.P. Binzel (eds.), University of Arizona Press, Tucson, USA, pp. 379–394.

Nesvorný D. et al. (2005) Evidence for asteroid space weathering from the Sloan Digital Sky Survey. *Icarus* 173, 132–152.

Ortiz J.L. et al. (2006) Detection of sporadic impact flashes on the Moon: Implications for the luminous efficiency of hypervelocity impacts and derived terrestrial impact rates. *Icarus* 184, 319–326.

Ortiz J.L. et al. (2012) Albedo and atmospheric constraints of dwarf planet Makemake from a stellar occultation. *Nature* 491, 566–569.

Pieters C.M. and L.A. McFadden (1994) Meteorite and asteroid reflectance spectroscopy: Clues to early solar system processes. *Annual Review of Earth and Planetary Sciences* 22, 457–497.

Popova O. et al. (2011) Very low strengths of interplanetary meteoroids and small asteroids. *Meteoritics and Planetary Science* 46, 1525–1550.

Sicardy B. et al. (2011) A Pluto-like radius and a high albedo for the dwarf planet Eris from an occultation. *Nature* 478, 493–496.

Tanbakouei S., J.M. Trigo-Rodríguez, J. Sort, P. Michel, J. Blum, T. Nakamura and I. Williams (2019) Mechanical properties of particles from the surface of asteroid 25143 Itokawa. *Astronomy & Astrophysics* 669, A119, doi: https://doi.org/https://doi.org/10.1051/0004-6361/201935380, 5 pp.

Trigo-Rodríguez J.M., and I.P. Williams (2017) Dynamic Sources of Contemporary Hazard from Meteoroids and Small Asteroids. In *Assessment and mitigation of asteroid impact hazards*, Springer, Astrophysics and Space Science, Vol. 46, pp. 11-32.

Trigo-Rodríguez J.M. and J.M. Madiedo (2013) *Asteroids and comets as progenitor bodies of meteorites: Their danger of impact with the Earth.* Revista Enseñanza de las Ciencias de la Tierra-AEPECT no. 21–3 (ISSN: 1132–9157), Asociación Española para la Enseñanza de las Ciencias de la Tierra, Madrid, pp. 293–301.

Trigo-Rodríguez J.M. et al. (2007) Asteroid 2002NY40 as a source of meteorite-dropping bolides. *Monthly Notices of the Royal Astronomical Society* 382, 1933–1939.

Trigo-Rodríguez J.M. et al. (2014) UV to far-IR reflectance spectra of carbonaceous chondrites – I. Implications for remote characterization of dark primitive asteroids targeted by sample-return missions. *Monthly Notices of the Royal Astronomical Society* 437, 227–240.

Trigo-Rodríguez J.M., H. Palme and M. Gritsevich (2017) Barcelona Asteroid Day 2015: Revisiting the threat by asteroid and comet impact. In *Assessment and Mitigation of Asteroid Impact Hazards, Astrophysics and Space Science* 46, Springer, pp. 1–9.

Urakawa S. et al. (2014) Fast rotation of a subkilometer-sized Near-Earth Object 2011 XA3. *The Astronomical Journal* 147, A121, 8 pp.

Vernazza P. et al. (2009) Solar wind as the origin of rapid reddening of asteroid surfaces. *Nature* 458, 993–995.

Walsh, K. J. and A. Morbidelli (2011). The effect of an early planetesimal-driven migration of the giant planets on terrestrial planet formation. *Astronomy and Astrophysics* 526, 126.

Chapter 6
Craters and Extinctions in the Geological Record

Understanding that the Earth is not, and never has been, oblivious to its cosmic environment, and has suffered countless collisions with other bodies throughout its history, could be described as one of the great adventures in human knowledge. Indeed, recognising this latent danger has not been an easy path and, as with other scientific disciplines, it is not yet widely known to the general public. Precisely for this reason, and writing as one of those responsible for the international initiative called *Asteroid Day,* we have been advocating the need to increase the visibility of the problem we are facing. As a result of this international effort coordinated by scientists, visionaries and enthusiasts of space exploration, in 2016 the United Nations decreed every 30 June to be Asteroid Day, in memory of the fall on Tunguska on that day in 1908.

However, the evidence of the danger is undeniably overwhelming. The geological and palaeontological data collected in the last decades, together with studies of cratering of the Moon and planets, as well as the dating and study of lunar rocks and meteorites, leave no room for doubt. Impacts are much more frequent than we would like. Moreover, the evidence even reveals that the Earth initially formed in multiple collisions between planetary embryos. Evidence accumulated in metallic meteorites, the cores of disintegrated planetary bodies, indicates that catastrophic fragmentations abounded, some of which delivered enough energy to evaporate and disintegrate these embryos with diameters thousands of kilometres across. These cascading fragmentations, with the help of dynamic resonances, are also the cause of the meteorites that continuously arrive on Earth (Jewitt,2008; Trigo-Rodríguez et al., 2015).

One of the greatest such catastrophes we know of is very close to us: our planet. Some 4.5 billion years ago an encounter occurred that, although it did not destroy our planet, would forever mark its evolution and that of the living beings that would inhabit it. It all came about as a result of a chance encounter between the proto-Earth and another embryo that has been agreed to be called Teia. As a result of this colossal encounter, a disc of ejected material was formed, which later, by progressive aggregation of the fragments, would give rise to the Moon. The dating of lunar

J. M. Trigo-Rodríguez, *Asteroid Impact Risk*, Impact Studies,
https://doi.org/10.1007/978-3-030-95124-5_6

Fig. 6.1 A huge rocky block excavated by an impact next to astronaut Harrison Schmitt during the exploration of the Apollo 17 Moon landing site in the Taurus–Littrow region (NASA)

meteorites and, in particular, the study of the oxygen isotope ratios of both planetary bodies proves it: the Moon and the Earth share a common origin.

In addition, exploration probes to the Moon and many other corners of our planetary system reveal that impacts were frequent, the further back in time they went (Baldwin, 1949; Shoemaker, 1998). The rocky planets were affected by large impacts and show undeniable traces. The evidence is compelling: a large impact must have reversed the rotation of Venus, while highly energetic impacts eroded the crust and part of the mantle of Mercury. Meanwhile, exploratory missions such as NASA's Dawn have revealed that the large asteroids in the so-called main belt between Mars and Jupiter were hammered by countless projectiles that created large layers of fractured materials, known as megaregolith, also covering the surface of the Moon (Fig. 6.1).

It was during the first hundred million years that the terrestrial planets, still in embryonic form, suffered the most intense bombardment, followed by a relative calm, broken only sporadically by a less frequent impact. This frequency continued to decline until about 3.8 billion years ago, when the migration of the giant planets Jupiter and Saturn caused a gravitational dispersion of the objects that formed part of the then highly populated outer asteroid belt, some of which were propelled towards the terrestrial planets (Trigo-Rodríguez, 2012a). The nearly 400 kg of rocks brought back by the Apollo astronauts revealed that the formation of most of the large craters and some lunar basins occurred during that brief period in Earth's evolution that we call the Late Heavy Bombardment (Fig. 6.2).

Some 4.56 billion years have passed since the creation of planetesimals, and since most of them have integrated to form planets, large collisions with objects hundreds

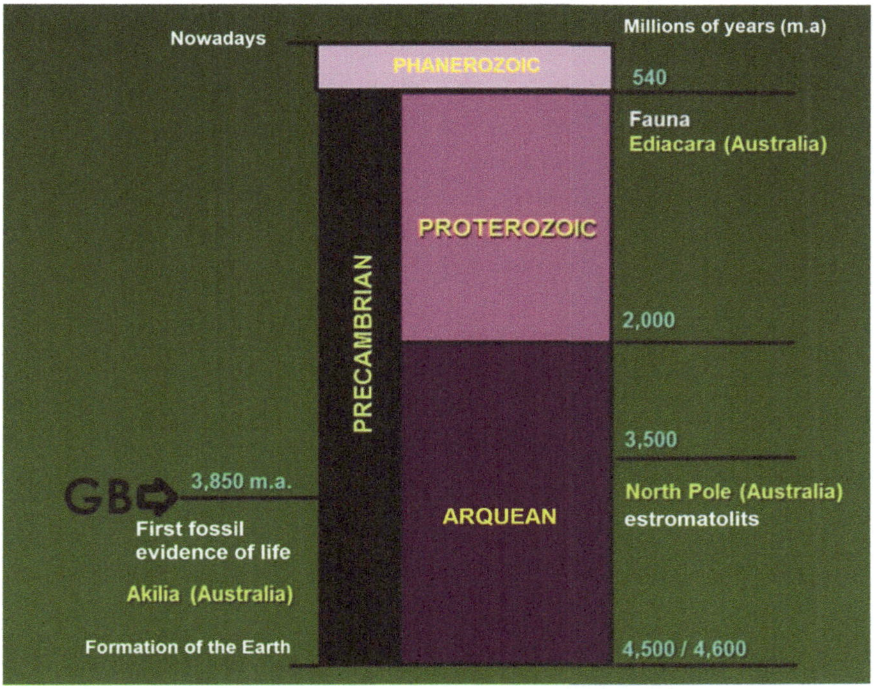

Fig. 6.2 The main geological eras and the Late Heavy Bombardment (GB) period

of kilometres in size no longer occur because the orbits of the largest asteroids are stable. This does not detract from the fact that there may have been a much rarer encounter, hundreds of millions of years ago (Ma), with a giant comet such as Hale–Bopp. However, potentially devastating impacts by asteroids of a few kilometres in diameter occur at frequencies of tens of Ma (see previous chapter). Certainly, for human life this is a huge scale but, given the stochastic nature of collisions and subsequent fragmentations in the main asteroid belt, it is not out of the question in the medium term. I will discuss this further below.

Craters: Direct Evidence of Colossal Impacts

From time to time the Earth has been subject to major collisions with kamikaze asteroids and comets. This is best evidenced by the 188 impact structures recognised by the *Earth Impact Database* (EID, 2015). In this chapter I will show how the dating of these craters allows us to quantify the frequency of such impacts based on geological evidence. However, the apparent scarcity of craters on the surface of our planet compared with that of the Moon arises because of the presence of atmosphere and oceans as well as the widespread

Table 6.1 Confirmed impacts in the Earth Impact Database that are older than 1 billion years. Note that there is no evidence of impacts earlier than about 2,500 Ma because they have been erased by tectonic processes (EID, 2015)

Name	Country	Age (Ma)	Diameter approx. (km)
Suavjärvi	Russia	~2,400	
Vredefort	South Africa	2,023 ± 4	160
Yarrabubba	Australia	~2,000	30
Sudbury	Canada	1,850 ± 3	130
Paasselkä	Finland	<1,0.800	
Keurusselkä	Finland	<1,800	30
Dhala	India	1,900 ± 200	
Amelia Creek	Australia	1,120 ± 520	
Shoemaker (Teague)	Australia	1,630 ± 5	30
Matt Wilson	Australia	1,402 ± 440	7.5
Goyder	Australia	<1,400	
Santa Fe	USA	<1,200	6–13
Iso-Naakkima	Finland	>1,000	
Lumparn	Finland	1,000	

development of plate tectonics and erosional processes, which hinder crater formation and smooth the resulting structures. Therefore, our record is biased and not without long-term gaps, although quite complete and convincing for the last few hundred million years (Table 6.1). Obviously, erosion hardly affects the most recent craters, which are shown to be perfectly bounded, but erosion by the elements causes crater walls to crumble and blend in with their surroundings. In fact, evidence of many craters can only be discovered by scanning images taken from space.

To get an idea of the bias that erosion and tectonic activity induces on the preservation of craters on the Earth's surface, Table 6.1 shows a compilation of the oldest impact structures excavated in the last billion years. Fortunately, we have the lunar record that informs us of the frequency of large impacts on the Earth–Moon system. Among the geologists who first realised the importance of identifying craters on Earth and comparing the flow of bodies on the Earth and the Moon was geologist Eugene Shoemaker (1928–1997) who completed a memorable compendium before suffering a fatal accident (Shoemaker, 1998). Indeed, all these studies are based on the fact that, as the Moon accompanies the Earth, a proportionality factor can be applied that takes into account its gravitational cross-section compared to the Earth's. Since the Moon has no atmosphere, and erosional processes are much slower, its surface is a fossil record of the most intense impact phases experienced by the Earth–Moon system. This allows us to quantify the flux of large bodies colliding with our planet.

There are certainly more ways to identify late impact traces, even if the crater has disappeared. Impacts not only excavate craters, but also generate rocks altered by their effects called impactites, among which we find tektites. Furthermore, it is suspected that they may be involved in the formation of some important metal deposits and even oil reserves (Ormö and Oms, 2013). In the following section, I will outline some of the detective work that can be followed to find clues to these processes, a real adventure with today's cutting-edge knowledge. The resulting discoveries are key to obtaining evidence of the direct and indirect effects of impacts on the Earth's history (Chapman and Morrison, 1994).

6.1 The Geological Scale: Looking for Traces of Catastrophes

As described above, the Earth is a geologically active planet and externally dominated by plate tectonics that regenerate its surface, so it is difficult to find evidence of impact structures older than a billion years. Even so, there are honourable exceptions which indicate that asteroids of tens or even hundreds of kilometres in size still reach the Earth (Table 6.1). However, large craters with a central collapse diameter greater than a hundred kilometres are extremely rare. They account for only 1% of the total, and in fact there are only three: Vredefort, Sudbury and Chicxulub. The latter will probably be familiar to us as it is the one associated with the extinction of the dinosaurs 65 million years ago.

Geological events are often divided into eons (Fig. 6.2). The Hadean eon spans from the beginning of our planet until 4,000 Ma ago. It is followed by the Archaean eon, which runs until about 2,500 Ma after the present. This is followed by the Proterozoic eon, which ends about 540 million years ago. Finally, the latter leads to the Phanerozoic eon in which we find ourselves (Fig. 6.3). In each of these eons, geological eras and periods have been defined, mainly on the basis of the fossil evidence that exists from those times. As is reasonable to imagine, as we move down the geological scale the evidence becomes more and more sparse. Thus, although there are methods of accurately dating the age of rocks, it becomes difficult to distinguish clear periods, particularly in the Archaean eon. It is also obvious to note that these rocks have often undergone significant metamorphism due to subduction and therefore lack the fossils that would allow such clear patterns of more recent eras to be established.

Fig. 6.3. The geological time eras of the Earth.

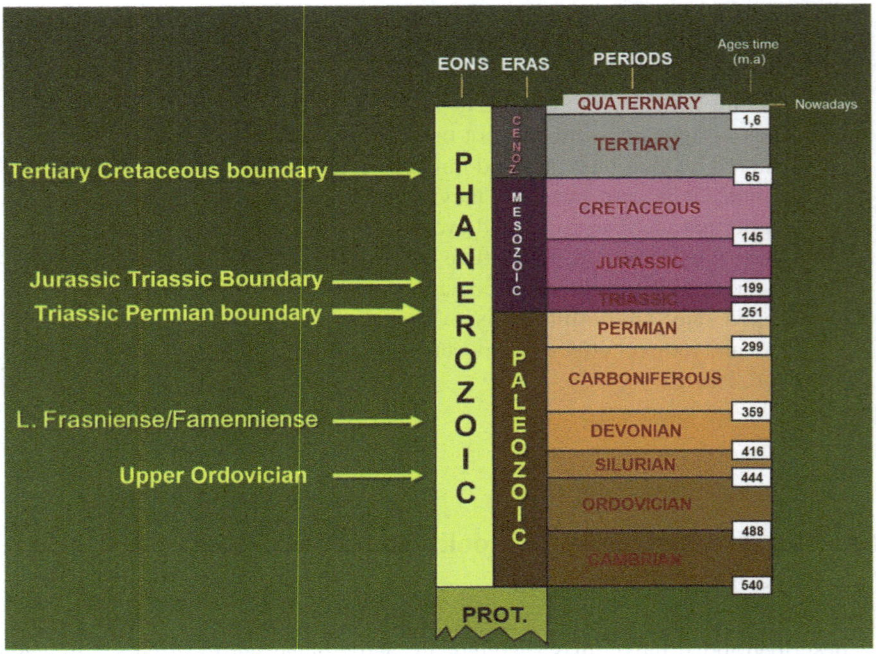

Fig. 6.3 Major sudden extinction events in the Phanerozoic eon

Devastating impacts are often mistakenly thought to be the stuff of science fiction, or perhaps the stuff of rambling in our collective imagination. However, the geological record indicates without a doubt that planetary-scale catastrophes have occurred, albeit fortunately infrequently. Much more biased is the record we have of impacts with regional devastation, as these may have been phenomena misunderstood by our ancestors. It is even possible that something with the destructive capacity of large superbolides would have been associated with divine punishment. I will return to these phenomena later.

These hecatombs are generally referred to as *mass extinctions*. They have been defined on the basis of stratigraphic legacy which, by preserving the fossils of organisms, clearly shows how populations vary at a single site as a function of time. Such variations when evidenced on a global scale, when found in multiple sites, show clear evidence of extinctions. Kauffman (1988) defined a mass extinction to have occurred when evidence was found for the extinction of a minimum of 50% of living species on a timescale between 1 and 3.5 Ma. This condition has since been moderated by Carlisle (1995), who redefined it to mean the extinction of at least 10% of species on a timescale equal to or less than 1 Ma.

In recent decades there has been a debate about the nature of such extinctions. While it is true that some of them may be due to endogenous climatic changes, induced for example by volcanism, others clearly point to their exogenous nature.

However, it should not be forgotten that the impacts have a whole phenomenology and unique products associated with them that can be identified without a doubt. I will return to all this evidence and list it below for the sake of argument. A classic and good review of this topic was by Koeberl (2001).

6.2 Phanerozoic Extinctions

The various extinctions that have occurred throughout the geological record are well known in the scientific literature but are rarely publicised, which is necessary to raise awareness of humanity's real enemies. The further back in time we go, the more difficult it may be to find a single explanation for a mass extinction. As the Earth's surface is subject to continuous regeneration due to plate tectonics and erosion itself, this should come as no surprise. We could say that the Earth, unlike other inactive rocky stars such as the Moon, is a planet that is erasing its wounds.

Here I will focus on the Phanerozoic eon, which is the most recent and therefore the best known from multiple studies. We will be familiar with both its eras and the periods into which it is subdivided (see Fig. 6.3). Different processes have been proposed to account for such mass extinctions, from those of biotic origin, such as epidemics or competition between species, to those of climatic or environmental origin, such as anoxia phenomena in the marine environment or variations in sea level due to the movement of tectonic plates that can also generate volcanic activity or relatively sudden climatic changes. However, in recent decades, the involvement of other exogenous phenomena, such as supernova explosions or the impact of large asteroids and comets, has been proposed and studied.

The explosion of a supernova in near-solar system environments is not a frequent event and this was proposed by Russell and Tucker (1971) as the cause of the extinction of the dinosaurs. The intrinsic problem with these processes is the deleterious effect of a flux of ionising radiation reaching a biosphere such as the terrestrial one. A bath in a dense X-ray or gamma-ray flux would in itself have a devastating effect on all living organisms because of its ability to destroy genetic heritage at the cellular level. While such starbursts must have affected evolution at some point, it is difficult to pinpoint how and when this happens because it leaves no discernible traces in the fossil legacy, unlike impacts with asteroids or comets.

Therefore, following the aim of this book, I will concentrate on the hazard associated with impacts and the existing evidence of such processes. Our view and interpretation are biased, as these are better represented for more recent geological scales. Five major terrestrial hecatombs have been identified on the basis of species extinctions in the stratigraphic legacy of the last 600 Ma during the Phanerozoic eon (Fig. 6.3). This does not detract from the fact that there have been others as or more important in the past, but these are less well known as the stratigraphic legacy becomes more biased further back in time, and for the older ones no fossil evidence is available (Shoemaker, 1998). Within the Palaeozoic, the extinctions occurred in the early Silurian (about 435 Ma ago) and in the late Devonian (360 Ma) and Permian

(245 Ma). To be precise, in the late Permian stratigraphic legacy, a sulphur enrichment has been discovered that is suggested to be associated with a large impact (Kaiho et al., 2001). Later, in the Mesozoic, two other mass extinctions occurred in the late Triassic (200 Ma) and Cretaceous (65 Ma). I will discuss the latter later, since it is thought to be associated with the extinction of the dinosaurs.

6.3 The Forerunners of Impact Theory

We could say that one of the first authors to reasonably consider the problem of impact with asteroids and comets was Isaac Newton (1672–1726/7), who in his youth (specifically in 1694) made a series of reasoned reflections on the impact of comets. Another clear example is the Frenchman Pierre-Louis Maupertuis (1698–1759), who in 1742 stated categorically that comets must have hit the Earth causing extinctions due to their enormous velocities.

The German meteorologist Alfred Wegener (1880–1930) was a pioneer in proposing that the craters on the Moon had been produced by impacts throughout the evolution of the solar system (Wegener, 1921). At that time there was also an important debate about the origin of a crater in Arizona, which Daniel Moreau Barringer (1860–1929) began to study in order to demonstrate that it was a structure excavated by a meteorite impact. We will return to that case in a later section, given its relevance to the geophysical community's belief in the importance of this natural process. Later, Ralph Baldwin (1912–2010) wrote a book on the origin of craters on the Moon and the importance of their scientific study (Baldwin, 1949).

Several more recent authors have also emphasised the danger of impacts. For example, Max Walker de Laubenfels (1894–1960) explained the extinction of the dinosaurs by an event similar to the one that caused the explosion of the Tunguska bolide, namely an *atmospheric outburst* as described in previous chapters. Later, in 1970, Canadian geologist Digby J. McLaren (1919–2004) claimed that a meteorite caused the mass extinction that took place some 365 million years ago (Upper Devonian). An excellent review of all this evidence is found in a later work of his (McLaren and Goodfellow, 1990).

Some geologists have also proposed impacts to explain the evidence associated with large extinctions in the palaeontological record. However, this narrative needs to explain in detail how the famous Arizona *meteor crater* was key in recognising the magnitude of our planet's impact with extraterrestrial objects.

The Scientific Adventure of The Meteor Crater

Arizona has some really curious geological structures. One of them is particularly so because of its crater-like shape and its approximately 1.2 km diameter. It was the American geologist Grove K. Gilbert (1843–1918) who studied it

Fig. 6.4 The two lithologies present in the Chelyabinsk meteorite; the upper darker one was created by the compaction of minerals under the colossal impacts suffered between asteroids before their arrival on Earth (CSIC-IEEC)

in 1891 when it was known as *Coon Butte.* Although a strong advocate of the impact origin of lunar craters, Gilbert came to the conclusion that the crater must have been produced by an explosion of volcanic vapour. Such a conclusion is still surprising to many, given that metallic meteorites around the crater were already common at the time. However, Gilbert believed that a meteorite of the magnitude needed to excavate such a crater and be buried inside it must have generated magnetic anomalies. He also thought that the volume of the crater generated, including the meteorite, should be larger than that of the crater walls and thus ruled out its meteorite origin.

Our particular adventure into understanding about the role of impacts in the history of the Earth continued in 1902, when the entrepreneur, miner and geologist Daniel Moreau Barringer learned of the existence of this curious structure. He immediately correlated the existence of the crater and the hundreds of metallic meteorites found around it with an impact, and thought he had in front of him a possible source of fame and business (Barringer, 1906). In fact, his first calculations led him to believe that at the bottom of the crater he should find a metallic meteorite weighing some 10 million tonnes. Following this, he spent 27 years devoting all his resources to this endeavour, a veritable fortune at the time, funding multiple drillings to find the native iron in the meteorite that he thought would make him a multimillionaire and famous. However, shortly before his death, the astronomer Forest Ray Moulton (1872–1952) calculated that most of the projectile's mass must have been vaporised during the excavation of the crater. Barringer's meteoric epic was reduced to a mirage. Nevertheless, the recognition of his work in identifying it as a meteor crater and promoting its study lasted until the mid-twentieth century, when the Meteoritical Society named it 'Barringer Meteor Crater'.

However, the debate was still raging and many geologists debated its meteoritic origin at the time. Perhaps it was the magnitude of the challenge to confirm such an origin once and for all that motivated Harvey H. Nininger (1887–1986), a renowned meteorite expert, to compile evidence of explosive excavation of the crater by a meteorite. At that time the well-known inventor Loring Coes (1812–1906) had discovered that silica under very high pressures produced a high-pressure variety. This polymorph of silica was named *coesite* in his honour and was to become the definitive proof they were looking for. However, it was the American geologist Eugene M. Shoemaker who made a detailed study of the crater and identified quartz-rich rocks with the idea of finally putting the question to rest. The detailed study of such samples at the United States Geological Survey (USGS) would reveal that they contained abundant coesite. They had found the first criterion for identifying an impact crater: recognising high-pressure minerals generated during its abrupt excavation. Later, they added other criteria such as so-called shatter cones, planar deformation features in minerals, or the presence of markedly meteoritic chemical elements implanted in the impacted rocks. Such criteria allowed the unambiguous identification of other impact structures around the world, and Shoemaker was honoured for this important contribution.

Today we know that the extremely high pressures and temperatures at which shock waves generated by impacts compress rocks cause irreversible changes in the composition and structure of minerals. Such processes transform the original minerals into high-pressure polymorphs, such as the former quartz to coesite or stishovite, graphite to diamond or zircon to reidite, among others (see Koeberl, 2001). Shock waves propagate through the material and deform it during the brief overpressure peak to produce these new *shock minerals* at pressures between 10 and 50 GPa. In addition, the higher shock can sometimes exceed 60 GPa, so it also dissociates the minerals, briefly melts them and generates mixed phases and glasses. An example would be silica glass or lechatelierite. An example of such phases is found in impactites but also in highly shocked meteorites like Chelyabinsk that have melt veins and mixed minerals, altering the physical properties (Fig. 6.4).

6.4 The Recognition of the Chicxulub Catastrophe and the k/t Boundary

The renowned American scientist Harold C. Urey (1893–1981), winner of the Nobel Prize in Chemistry in 1934, was convinced of the important role of impacts in the history of the Earth. However, among his many contributions, this aspect is not so well known. His research on this subject was compiled in an article he submitted to the prestigious journal *Nature*, in which he suggested that several extinctions in the last 50 Ma could be the result of impacts by large comets (Urey, 1973). The enormous kinetic energy of these fast-moving objects and their haphazard steps were already seen by Urey as a cause for concern.

But how could the importance of impacts in the recent history of planet Earth be demonstrated? Here, the accumulated knowledge from the study of the fossil record compiled by hundreds of authors has played a key role. I am going to focus on a particular moment in Earth's history between the Cretaceous and Tertiary periods, known in the literature as the K/T boundary. Its acronym comes from the German *Kreide/Tertiär* and it marks the boundary between the Cretaceous and the Tertiary. Following this brief interval, the large saurians that had dominated the planet until then disappeared. The level of diversity and wide distribution of dinosaurs has been well exemplified in the fossil record, so their relatively sudden extinction raised many questions among scholars. Multiple hypotheses have been proposed to explain the causes of such a debacle for the then dominant species: how was it possible for species of such magnitude and number, widely distributed across the planet, to experience such a sudden decline?

It was the American physicist Luis W. Álvarez (1911–1988) and his son, the geologist Walter Álvarez, who published the most important discovery on the impact hypothesis (Álvarez et al., 1980). The answer they provided came from direct evidence, from a real impact marker, in the geological record. In fact, that Alvarez father and son were a physicist and a geologist was the perfect interdisciplinary perspective, given the nature of the evidence found. The first evidence came from the detection of iridium (Ir) contained in a clay stratum distributed across our planet. Although its presence must be global, it outcrops only in certain regions and acts as a marker of the so-called K/T boundary that marks the end of the Mesozoic era and the beginning of the Cenozoic era. Why was the high abundance of iridium detected in this K/T stratum key? The answer comes from the fact that enrichment in elements which are scarce in the Earth's crust acts as a *tracer* of the event itself. Since certain metals travelled into the Earth's interior during the primordial differentiation of our planet, they can only be found in strata that must have been produced through an exogenous process, such as an impact. An asteroid before impact may lose most of its mass in the fireball, and the impact itself also generates dust and impactites that are thrown enormous distances depending on the energy of the onslaught (Hildebrand, 1993). The projectile materials are mostly vaporised and thrown into the atmosphere where they then reside for weeks or months before falling to the surface. In fact, we know that asteroids smaller than about a hundred metres fracture efficiently in the atmosphere and, therefore, craters of the dimensions produced by these objects are rare (Hills and Goda, 1993; Bland and Artemieva, 2003). Therefore, certain elements that are relatively abundant in meteorites such as those of the platinum group (Pt, Ir, Ru, Rh, Pd and Os) are excellent markers of an impact with an extraterrestrial object (Álvarez et al., 1980).

The discovery of iridium in this well-defined stratum of limestone rocks across the globe occurred several years before the discovery of the Chicxulub crater at the bottom of the Gulf of Mexico. Chicxulub is a partially hidden crater, in this case under the sea, which would only be detected in the Gulf of Mexico after a detailed gravimetric survey for resources—see Fig. 6.5. The measured Ir values were

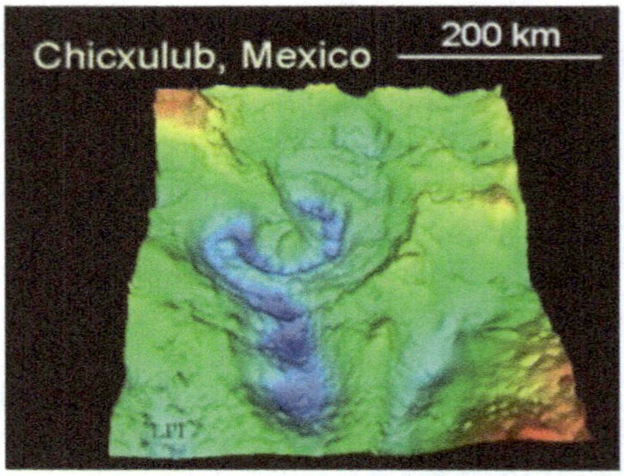

Fig. 6.5 A detailed gravimetric survey made to search for resources discovered the concentric walls of the Chicxulub impact (Lunar and Planetary Institute)

unequivocally indicative of an impact, as the Alvarez found iridium concentrations in three widely spaced deposits (Denmark, Italy and New Zealand) that were 160, 30 and 20 times higher than normal at the time. The Alvarez paper in *Science* became a must-read because it calculated that the asteroid should have had a diameter of 10 ± 4 km and predicted an impact crater of hundreds of km, but also proposed a number of post-impact effects that should not be neglected. Impact with the asteroid at a typical velocity from the main asteroid belt would not only have created a crater hundreds of kilometres across, but also injected some of its mass and excavated terrestrial material into the atmosphere in the form of fine dust rich in these platinum-group elements (Kyte et al., 1981). It also predicted an abrupt cooling phase due to this accumulated dust in the stratosphere (something already observed as a consequence of volcanism). In addition to this evidence, the Chicxulub impact had associated climatically active gases, particularly from the injection of sulphur from the plume (Pierazzo et al., 2003). It is now accepted that this major impact, followed by all these lethal consequences, caused the disappearance of the dinosaurs at the end of the Cretaceous period.

The origin of the asteroid that produced the Chicxulub crater is still a matter of debate and will fill future research. One scenario proposed by Bottke et al. (2007) suggested the fragmentation of an asteroid family known by its largest member (298) Baptistina could have led to destabilisation of the projectile that generated the impact crater. However, spectroscopic follow-up indicated that the Baptistina family has an S spectral class, rather than an Xc-type composition, making it an unlikely source of such a projectile, which had a carbonaceous chondritic composition (Kyte, 1998; Trinquier et al., 2006). Subsequent studies of the dimensions and distribution of the Baptistina family suggest that it may have broken up later, making the cosmic carom even more difficult. The paths followed by the asteroids depend very much on the dynamical processes I mentioned earlier, called resonances. They are far from instantaneous, requiring tens and sometimes hundreds of millions of years for an asteroid to be efficiently launched towards Earth. The absence of a clear candidate has been exploited in a recent paper by Siraj and Loeb (2021) to propose that the projectile could have come from the fragmentation of a long-period comet after a

Fig. 6.6 Artist's rendering of the impact that excavated the Chicxulub crater (NASA)

close passage to the Sun. There is no doubt that our understanding of comets has changed enormously and that some of the objects scattered into high-eccentricity orbits during the migration of the giant planets may have been transitional asteroids, halfway between a comet and an asteroid. Hopefully, new evidence will allow us to pinpoint the nature and mechanism that launched such a colossal projectile to cause such an era change on our planet.

6.5 The Impact of Asteroid Eltanin

Despite what we may think, we don't have to go that far back in time to find evidence of a collision of our planet with a kilometre-sized asteroid. Exploration of the deep ocean has led to great discoveries, and in this field it was not to be outdone. A significant part of this progress was made by understanding oblique impacts in water and identifying the main differences with direct impacts excavating craters (Pierazzo and Melosh, 2000).

In 1965 the icebreaker USNS Eltanin collected sediments on the ocean floor (at a depth of about 5 km) that contained a suspected iridium enrichment (Kyte et al., 1981). Furthermore, the sediments recovered by that ship were Eocene and Palaeocene but chaotically mixed and deposited on the sea floor, and contained meteorite fragments (Kyte et al., 1988; Gersonde et al., 1997). These authors estimated that the asteroid had a diameter between 1 and 4 km and an explosive power of approximately 10 million megatonnes of TNT. This case exemplifies the extent to which we are unaware of the impacts that may have occurred on the ocean floor, although exploration of the deep ocean provides new evidence every day.

These studies revealed that the Eltanin impact had occurred in the South Pacific Ocean at the beginning of the Quaternary period about 2.5 Ma ago, in the Pliocene.

Gersonde et al. (1997) further explained that certain sediments with stacked fossils have been found off the coasts of Peru and East Antarctica. This was not difficult to explain, considering that the impact of an asteroid of such dimensions must have caused tsunamis with waves of tens of metres in amplitude that could have reached certain continental margins with heights amplified by an order of magnitude. Computer modelling by Steven N. Ward has made it possible to reconstruct the extent of such mega-tsunamis (Fig. 6.7).

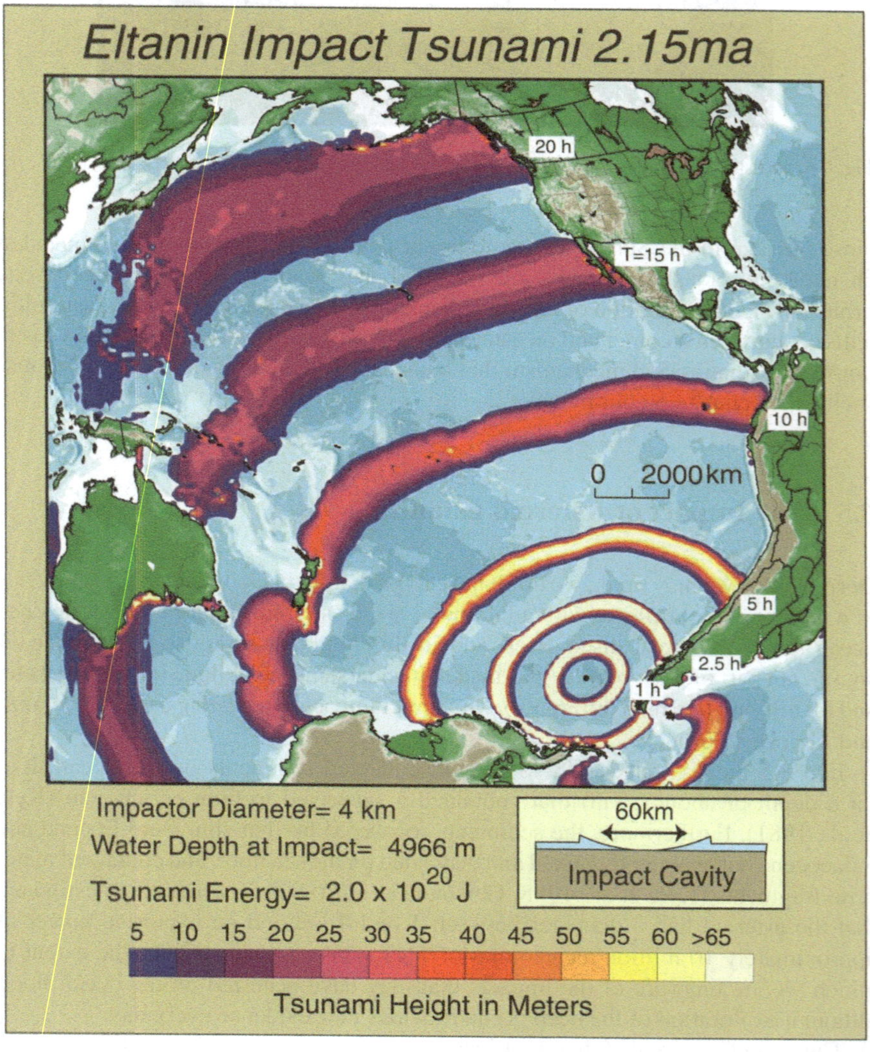

Fig. 6.7 The tsunami generated by the impact of the asteroid Eltanin (courtesy of Steven N. Ward)

It is important to ask about the implications of such impacts on the ocean, as authors such as Chapman and Morrison (1994) have already done. Importantly, these impacts could be very detrimental to the terrestrial biosphere because of their ability to drive large amounts of water vapour and salts into the upper layers of the Earth's atmosphere. Such products could be insulating, poisonous to aerobic respiration and even nitrous oxide could be capable of destroying the ozone layer (Toon et al., 1997).

There is no doubt that the Eltanin impact is evidence of a large event in a relatively recent period. It may well be that exploration of the deep ocean will provide further evidence of these impacts. A detailed compilation of the magnitude and extent of the ecosystem implications of a collision was made by McLaren and Goodfellow (1990). This work provides a good reminder that this danger is real, has the potential to reach global dimensions and cannot be ignored. Our difference from the dinosaurs is that we have learned to read the messages sculpted into the rocks—but are we dedicating enough research to react in time to defend life from the next big impact?

Questions for Chapter 6

Q.1. This graph represents in logarithmic scale the flux of asteroids and meteoroids reaching the Earth according to different types of observations. The lower abscissa axis provides the kinetic energy of the body (given in kT of TNT), while the upper axis gives the diameter of the body in metres. The ordinates represent the impact frequency in an accumulated way: impacts per year. The red pentagon can be used to learn how to read the graph. It marks the frequency of impact for km-sized bodies that can be read in the ordinates: 10^{-6}, or one impact every one million years. Calculate the impact frequency for bodies of the approximate size of the Tunguska event, assuming it was about 50 m in diameter. [Solution: We search in the logarithmic scale of the top abscissa axis where we look for the fifth mark between 10 and 100. Then we follow a vertical line until it crosses the diagonal flux data. At that crossing point we follow a horizontal line up to crossing the ordinates that provide an impact frequency of about 10^3. Statistically, an impact of such magnitude, 10,000 metric tonnes of TNT, occurs once per millennium.]

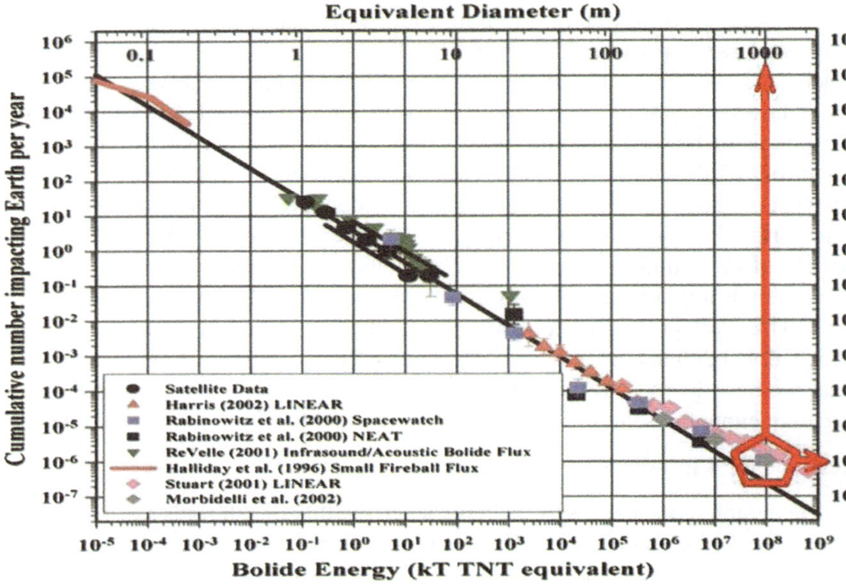

Q.2. Taking into account the averaged frequency of impacts of km-sized asteroids, and considering that there is only one impact crater confirmed to be less than 1 million years old, the 14-km-diameter Zhamanshin crater in Kazakhstan, should we be worried about the next large impact on our planet being overdue? [Solution: Certainly, we should not worry about it. On statistical grounds, the estimated impact frequencies are averaged values, but depend on stochastic processes so they reach our planet randomly.]

Q.3. On our planet the number of craters of a given diameter is influenced by erosion. In this graph from Hergarten et al. (2019) the blue line shows the number and ages of craters (20 km across or larger) found so far and the predicted distribution (green line) according to a simulation based on constant impact flux combined with erosion. How will you get a more precise measurement of the asteroid and comet flux over our planet over the eons? [Solution: We can measure the number of craters on the Moon, a planetary body linked to our planet. The graph shows that terrestrial erosion processes are biasing our understanding of the flux of terrestrial impactors, so to gain insight it is better to quantify the flux measured on the Moon, unaffected by such processes over long timescales.]

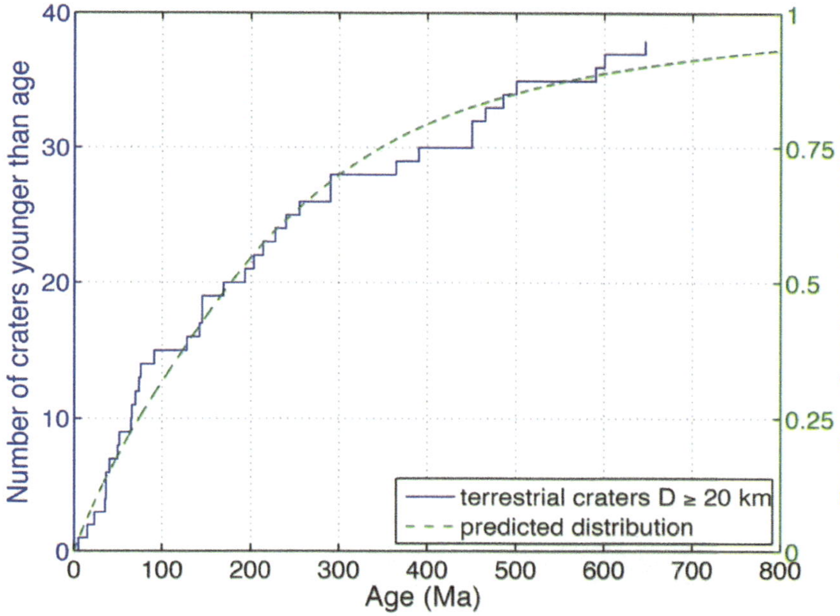

Q.4. Ocean-vaporizing impacts: by knowing that the mass of water forming the oceans is 1.4×10^{24} g and the latent heat of vaporization of water at 273 K is 2.5×10^{10} ergs/g and the specific heat of water vapor is 1.9×10^{7} ergs/g, try to answer these questions: (a) What is the impact energy required to vaporize the Earth's oceans? (b) What is the asteroid diameter for a typical collision velocity of 17 km/s? [Solution: First of all, remember that as a first approximation we consider an energy balance for the impact process as proposed by Sleep et al. (1989) in which the impact energy is distributed as follows: a quarter of the energy is spent in vaporizing water, while the rest is buried in the crust and mantle, or radiated to space. (a) A minimum of 4.4×10^{34} ergs are required for an ocean-vaporizing impact. (b) The asteroid should be about 450 km in diameter, similar in size to Vesta]

Q.5. Impacts excavating lunar basins: try to search in scientific literature for numerical relationships to compute the energy of the impact and the diameter of the craters. (a) How can we compute the size of a crater knowing the impact energy of an asteroid hitting the Moon? (b) How much energy is needed to excavate the Orientale Basin?

[Solution: (a) There are many semi-empirical approaches to compute the material excavated from large craters on planetary bodies without atmosphere. These are the result of simulations and impact experiments. Just as an example, here we follow the approach of Schmidt and Housen (1987). They used two scaling relations for the apparent volume and diameter of simple craters excavated in rock. In the equations appear the density of the impactor (ρ_i) and the crust (ρ_c), E is the impact energy, φ is the incidence angle, v is the impactor velocity, and g is the surface gravity. The ejected volume (V_e) excavated by the impact is given by equation [1], while the transient crater diameter (D_t) is given by equation number [2], improved by Zahnle and Sleep (2006) to take into account the angle of impact (Melosh, 1989). Take care to use compatible units to get a correct result.

To compute the final diameter of the complex crater (D) we have equation [3] related with the diameter of the transient crater (D_t) and the complex crater (D_c). The equations that we need to infer the size of our crater are:

$$V_e = 0.23(\rho_i/\rho_c)^{0.22} \times (E/\rho_c)^{0.78} \times v^{-0.26} \times g^{-0.65} \times \sin\varphi \quad (6.1)$$

$$D_t = 1.6(\rho_i/\rho_c)^{0.07} \times (E/\rho_c)^{0.26} \times v^{-0.08} \times g^{-0.22} \times (\sin\varphi)^{0.33}[\text{in cm}] \quad (6.2)$$

$$D = 1.2\, D_t^{1.13} \times D_c^{-0.13} \quad (6.3)$$

b) To excavate the Orientale Basin, an impact energy of about 2×10^{33} ergs is needed.

Figure for use with Q.5. The Orientale Basin is the youngest of the large lunar basins. The outer ring is about 950 km in diameter. (Credit: NASA/GSFC/University of Arizona.)

References

Álvarez L. et al. (1980) Extraterrestrial cause for the Cretaceous-Tertiary Extinction. *Science* 208, 1095–1108.

Baldwin R. (1949) *The Face of the Moon*. University of Chicago, USA, 239 pp.

Barringer D.M. (1906). Coon Mountain and its crater. *Proceedings of the Academy of Natural Science of Philadelphia* 57, 861–886.

Bland, P.A. and N.A. Artemieva (2003) Efficient disruption of small asteroids by Earth's atmosphere. *Nature* 424, 288–291.

Bottke, William F., D. Vokrouhlický, D. Nesvorný (2007) An asteroid breakup 160 Myr ago as the probable source of the K/T impactor. *Nature* 449(7158), 48–53. https://doi.org/10.1038/nature 0607010.1038/nature06070

Brown P., R.E. Spalding, D.O. ReVelle, E. Tagliaferri and S.P. Worden (2002) The flux of small near-Earth objects colliding with the Earth. *Nature* 420, 294–296.

Brownlee D. et al. (2006) Comet 81P/Wild 2 under a microscope. *Science* 314, 1711–1716.

Carlisle D.B. (1995) Dinosaurs, Diamonds, and Things from Outer Space, The Great Extinction, ISBN 0804723923, Cambridge University Press, Hardback, 1995.

Chapman C.R. and D. Morrison (1994) Impacts on the Earth by asteroids and comets: Assessing the hazard. *Nature* 367, 33–40.

EID (2015) Earth Impact Database: http://www.passc.net/EarthImpactDatabase

Ellis J. and D.N. Schramm (1995) Could a nearby supernova explosion have caused a mass extinction? *Proceedings of the National Academy of Sciences* 92(1), 235–238.

Gersonde R. et al. (1997) Geological record and reconstruction of the late Pliocene impact of the Eltanin asteroid in the Southern Ocean. *Nature* 390, 357–363.

Gomes R., H.F. Levison, K. Tsiganis and A. Morbidelli (2005) Origin of the cataclysmic Late Heavy Bombardment period of the terrestrial planets. *Nature* 435, 466–469.

Hergarten S., G. Wulf and T. Kenkmann (2019) Comment on 'Earth and Moon impact flux increased at the end of the Paleozoic'. *Science* 365, issue 6450, eaaw7471.

Hildebrand A.R. (1993) The Cretaceous/Tertiary boundary impact (or the dinosaurs didn't have a chance). *Journal of the Royal Astronomical Society of Canada* 87, 77–118.

Hills J.G. and M.P. Goda (1993) The fragmentation of small asteroids in the atmosphere. *The Astronomical Journal* 105(3), 1114–1144.

Jewitt D. (2008) Kuiper Belt and Comets: An observational perspective. In *Trans-Neptunian Objects and Comets*, Swiss Society for Astrophysics and Astronomy, Springer-Verlag, Heidelberg, Germany pp. 1–78.

Kaiho K. et al. (2001) End-Permian catastrophe by a bolide impact: Evidence of a gigantic release of sulfur from the mantle. *Geology* 29, 815–818.

Kauffman, E. G. (1988): The Dynamics of Marine Stepwise Mass Extinction. — In: Lamolda, M., Kauffman, E. G. & Walliser, O. H. (eds.): Paleontology and Evolution: Extinction Events. — *Rev. Española de Paleontología*., n. extrord., 57–71.

Koeberl K. (2001) Craters on the Moon from Galileo to Wegener: A short history of the impact hypothesis, and implications for the study of terrestrial impact craters. *Earth, Moon and Planets* 85–86, 209–224.

Kyte F.T. (1998) A meteorite from the Cretaceous/Tertiary boundary. *Nature* 396, 237–239.

Kyte F.T., Z. Zhou and J.T. Wasson (1981) High noble metal concentrations in a late Pliocene sediment. *Nature* 292, 417–420.

Kyte F.T., Z. Zhou and J.T. Wasson (1980) Siderophile-enriched sediments from the Cretaceous-Tertiary boundary. *Nature* 288, 651–656.
Kyte F.T., L. Zhou and J.T. Wasson (1988) New evidence on the size and possible effects of a late Pliocene oceanic impact. *Science* 241, 63–65.
McLaren D.J. and W.D. Goodfellow (1990) Geological and biological consequences of giant impacts. *Annual Review of Earth and Planetary Sciences* 18, 123–171.
Melosh H.J. (1989). Impact cratering: A geologic process. *Oxford Monographs on Geology and Geophysics*, No. 11. Oxford University Press, New York, USA, 245 pp.
Ormö J. and O. Oms (2013) The formation of impact craters. A fundamental process in the Solar System. *Revista Enseñanza de las Ciencias de la Tierra-AEPECT 21*, No. 3, 310–320.
Pati J.K. and W.U. Reimold (2007) Impact cratering: Fundamental process in geoscience and planetary science. *Journal of Earth System Science* 116, 81–98.
Pierazzo E. and H.J. Melosh (2000) Understanding oblique impacts from experiments, observations and models. *Annual Review of Earth and Planetary Sciences* 28, 141–167.
Pierazzo E., A.N. Hahmann and L.C. Sloan (2003) Chicxulub and climate: Radiative perturbations of impact-produced S-bearing gases. *Astrobiology* 3, 99–118.
Russell, D. and W. Tucker (1971) Supernovae and the extinction of the dinosaurs. *Nature* 229, 553–554.
Schmidt R.M. and Housen K.R. (1987) Some recent advances in the scaling of impact and explosion cratering. *International Journal of Mechanical Sciences* 5, 543–560.
Shoemaker E.M. (1998) Impact cratering through geologic time. *Journal of the Royal Astronomical Society of Canada* 92, 297–309.
Siraj, A. and A. Loeb (2021) Breakup of a long-period comet as the origin of the dinosaur extinction. *Scientific Reports* 11, 3803. https://doi.org/https://doi.org/10.1038/s41598-021-82320-2
Sleep N.H., K.J. Zahnle, J.F. Kasting, and H.J. Morowitz (1989) Annihilation of ecosystems by large asteroid impacts on the early Earth. Nature 342, 139-142.
Steel D. (1995) *Rogue Asteroids and Doomsday Comets*. Wiley, New York.
Toon, O.B., K. Zahnie, D. Morrison, R.P. Turco and C. Covey (1997) Environmental perturbances caused by the impacts of asteroids and comets. *Reviews of Geophysics* 35, 41–78.
Trigo-Rodríguez J.M. (2012a) *Las raíces cósmicas de la vida*. Colección El Espejo y la Lámpara, Servei Publicacions UAB, Barcelona, 241 pp.
Trigo-Rodríguez J.M. (2012b) Meteoritos. Editorial Catarata-CSIC, collection ¿Qué sabemos de?, Consejo Superior de Investigaciones Científicas, Madrid, 117 pp.
Trigo-Rodríguez J.M., E. Lyytinen, M. Gritsevich, M. Moreno-Ibáñez, W.F. Bottke, I. Williams, V. Lupovka, V. Dmitriev, T. Kohout and V. Grokhovsky (2015) Orbit and dynamic origin of the recently recovered Annama's H5 chondrite. *Monthly Notices of the Royal Astronomical Society* 449, 2119–2127.
Trinquier, A. et al. (2006) The nature of the KT impactor. A 54Cr reappraisal. *Earth and Planetary Science Letters* 241, 780–788.
Urey H.C. (1973) Cometary Collisions and Geological Periods. Nature 242, 32-33.
Wegener A. (1921) *Die Entstehung der Mondkrater*. Vieweg-Verlag, 48 pp.
Zahnle K. and N.H. Sleep (2006) Impacts and the early evolution of life. In *Comets and the Origin and Evolution of Life* by P.J. Thomas et al. (eds.), Springer, pp. 207–251.

The manufacturer's authorised representative in the EU is Springer Nature Customer Service Centre GmbH, Europaplatz 3, 69115 Heidelberg, Germany. If you have any concerns regarding our products, please contact ProductSafety@springernature.com

Printed and bound by CPI Group (UK) Ltd, Croydon, CR0 4YY
30/09/2026
02225259-0005